Rockhounding Montana

Help Us Keep This Guide Up to Date

Every effort has been made by the authors and editors to make this guide as accurate and useful as possible. However, many things can change after a guide is published—trails are rerouted, regulations change, techniques evolve, facilities come under new management, etc.

We would love to hear from you concerning your experiences with this guide and how you feel it could be improved and kept up to date. While we may not be able to respond to all comments and suggestions, we'll take them to heart, and we'll also make certain to share them with the authors. Please send your comments and suggestions to the following address:

> The Globe Pequot Press
> Reader Response/Editorial Department
> P.O. Box 480
> Guilford, CT 06437

Or you may e-mail us at:

> editorial@GlobePequot.com

Thanks for your input, and happy trails!

Rockhounding Montana

A Guide to all of Montana's Best Rockhounding Sites

Second Edition

Montana Hodges and Robert Feldman

GUILFORD, CONNECTICUT
HELENA, MONTANA
AN IMPRINT OF THE GLOBE PEQUOT PRESS

FALCONGUIDES®

Copyright © 2006 by Morris Book Publishing, LLC
A previous edition of this book was published by Falcon Publishing, Inc. in 1985.

ALL RIGHTS RESERVED. No part of this book may be reproduced or transmitted in any form
by any means, electronic or mechanical, including photocopying and recording, or by any
information storage and retrieval system, except as may be expressly permitted by the 1976
Copyright Act or by the publisher. Requests for permission should be made in writing to The
Globe Pequot Press, P.O. Box 480, Guilford, CT 06437.

Falcon and FalconGuides are registered trademarks of Morris Book Publishing, LLC.

All interior photographs are by Montana Hodges unless otherwise noted.
Maps created by Mapping Specialists © Morris Book Publishing, LLC

Library of Congress Cataloging-in-Publication Data

Hodges, Montana.
  Rockhounding Montana / Montana Hodges and Robert Feldman — 2nd ed.
      p. cm. — (A Falcon guide)
  Includes bibliographical references and index.
  ISBN 978-0-7627-3682-9

1. Rocks—Collection and preservation—Montana—Guidebooks. 2. Minerals—Collection and
preservation—Montana—Guidebooks. 3. Fossils—Collection and preservation—Montana—
Guidebooks. 4. Montana—Guidebooks. I. Feldman, Robert (Bob) II. Title. III. Falcon guide.
  QE445.M9F45 2006
  552.09786--dc22

                                           2006004811

Printed in the United States of America
Second Edition/Seventh Printing

For the author of the first edition of *Rockhounding Montana,* Robert Feldman, and his wonderful wife, Claudia. Had it not been for their years of hard work and dedication, many sites in Montana would remain silent.

Contents

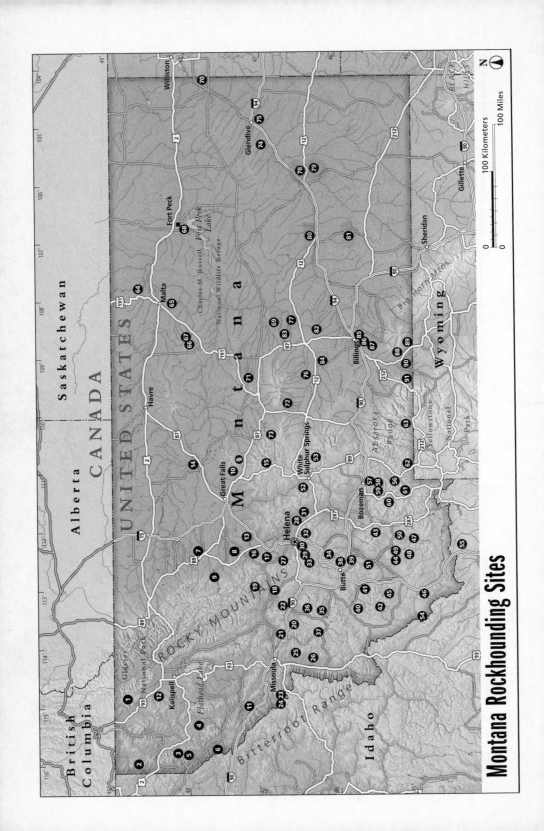

Montana Rockhounding Sites

Acknowledgments

First and foremost I would like to thank Mom and Dad; had they named me Virginia things would have been so different. A special thanks to Dad, who has inspired me to go outside and explore the natural world since I was a child. He brought science to my life and a love of open space to my soul. And a special thanks to Mom, who has given me the strength and ambition to follow my dreams and the attitude to make it all come true.

Thanks to the author of the first edition of *Rockhounding Montana,* Robert Feldman, who lent great help to the second edition and is responsible for the geologic integrity of this rockhounding guide. He really helped me keep my strata straight.

Also a due thanks to the man behind the curtain, Jimmy Goodman, who traveled with me throughout Montana and who shares the love of rockhounding. Without his expertise many fossils wouldn't have been identified and the road would have been far less traveled.

It is nearly impossible to properly express my gratitude to the geology department at Cal State Sacramento, particularly to the department chair, Dr. Dave Evans, who assured all the support a student in the field could ever dream of, not to mention the proper equipment. He was also responsible for the help of technician Steve Rounds, the geologist who spent much time identifying my many mysterious rocks and minerals. Thank you Steve, you rock.

I never cease to be amazed by the warmth and welcoming Montanans condone to their visitors. Never will there be a state so charming and honest. The warm welcome of the Travel Montana publicity tour is thanks to amazing publicist Ric Bourie (a special thanks to Ric, also, for inspiring me to learn to play the fiddle!) who developed the first state-sponsored rockhounding/journalism tour and Doug Smith of Missouri River Travel, who worked with Ric. With Doug the experience of Montana is even more authentic, and his love of eastern Montana inspires me still. Hopefully, some day I will be able to out-hike him.

A grand thanks to Rob and Libby Knotts, owners of the beautiful Gray's Coulee Guest Ranch in Lambert, who entertained all of us rockhounds so graciously and offered me the unique opportunity to live my dream of residing on an eastern Montana ranch, even if it was just for a couple of days!

Thanks, too, to Ken Rohn, writer for *Rock and Gem* magazine. Meeting Ken has affirmed that I am not crazy to decide to write about rocks, and his dedication has convinced me I can make a career out of it (my parents want to speak with you, Ken!).

Of course, the biggest thanks goes to all the rockhounds, without whom life would be so dull. Hopefully this book will bring many adventures and trips filled with happiness and, of course, rockhounds will want to pass it on to a friend. See you in the field!

—Montana Hodges

Introduction

Only one state could be known as the treasure state, and Montana holds that honor. Possibly no reference could be more suitable for a land of such rich history, vibrant culture, and physical beauty, not to mention so many great rocks! Montana is home to gold caches, ruby-spiked rivers, miners' tales of precious cobalt sapphires clogging sluice boxes, and dinosaur beds so vast that paleontologists can't dig them up as quickly as they erode from the prairies. Montana is a state with legends of the "richest hill on earth," copper kings, Custer's Last Stand, flashy minerals in mysterious ghost towns, fossils hunts in eerie "badlands," trails traversing through petrified forests, and the chance to uncover ancient seas. Montana embodies the Wild West, a rugged outback, an isolated prairie, a bustling culture, and the chance for an adventure in what may be a rockhound's paradise.

Tailings piles the size of mountains, like this one in the Neihart Mining district, stand as testimony to the mining craze in Montana's early history.

Montana's geography of today is best described by such words as land, space, mountains, and sky. Only three states—Alaska, Texas, and California—have an area bigger than Montana's 147,046 square miles. The name "Montana" is derived from the Spanish word for mountain, and one glance at a state map reveals dozens of ranges in the west. Some mountain-crest elevations rise above 12,000 feet. Most of the state's population and tourism can be found in the valleys that divide these western ranges. Nestled in these valleys are historical towns such as Helena, Missoula, Butte, and Bozeman.

The mountainous western third of Montana contains a treasure trove of precious gems and minerals. Suitably, the state's motto of *Oro y Plata* is Latin for "gold and silver." Deposits of many metals, including gold, silver, and copper, are located throughout the western mountains and set Montana's roots as a major mining state. Garnets, topaz, and sapphire (the state's precious gem) erode out of their many deposits in the west.

On the other side of the Continental Divide, the line at which water on the east flows to the Atlantic Ocean and water on the west flows to the Pacific, the landscape fades into hundreds of miles of rolling hills and prairie grasses. Here, the Great Plains and the Yellowstone River lay uninterrupted under infinite sky. Only eroded desert areas known as "badlands" stand as testament to the extremities of plains weather and eastern living. Arctic air from the north influences the weather of the plains. Low-lying mountains and prairie offer little shelter from the freezing wind of the winter and burning heat of the summer. But to an appreciative observer and the small percentage of the state's population who reside there, the lunar landscape is worth the setback. The mountains may be a source for the state's name, but the plains of eastern Montana are the inspiration for the state's most famous nickname—"Big Sky Country."

Meriwether Lewis and William Clark were the first recorded American explorers to officially travel across Montana during their U.S. government expedition of 1805–1806. The men represented the beginning of contact between the native tribes of Montana and the ambitious hunters to come. A goal of the expedition was to evaluate profit that could be made from land exploitation throughout the territory. At the time fur trapping was thought to be the most promising western enterprise. After Lewis and Clark spoke of the Montana territory as "richer in beaver than any other country on earth," trappers and traders flooded the region. The fur craze of the 1800s was the beginning of the end to the original lifestyle of Montana's Native Americans. Of course, it also didn't take long for Montana's newest inhabitants to discover metal deposits and begin the quest to remove the earth's minerals.

Eastern Montana is responsible for the state's nickname—Big Sky Country.

The fur craze, mining craze, and the new settlers pushed the Native Americans farther from the necessary lands for survival. The Sioux and Cheyenne fought the last successful battle by natives in 1876 at the Battle of Little Bighorn. As the desirable land became more scarce and after a series of bloody conflicts, the last native people were pushed onto reservations throughout the state. Today several reservations occupy Montana territory, and Native Americans continue to have a strong physical and cultural presence in the state.

Montana entered the union in 1889, becoming a young 41st state of the United States. The Montana of today is based less on mining natural resources and more on agriculture. Agriculture is one of the state's primary industries, but it is more struggling than robust. The population of the state has always managed to remain small, never topping more than one million people. The state may be one of the largest in area, but it ranks 44th in population.

Home to some of the nation's most popular national parks, from the ice-capped peaks of Glacier to the spurting hot springs of Yellowstone, Montana encompasses vast areas of protected lands. Crystal lakes, undisturbed forests, clean air, and the largest variety of mammals of any state are just a few of the traits that fulfill the desire for unspoiled landscape. A state with three times as many cows as people, it represents a different world inside a busy country. Montana truly is America's outback.

Montana's Geologic Introduction

The experience of rockhounding in Montana is greatly enhanced by an understanding of the state's geology. Most rockhounds aren't geologists, and most geologists don't write books on rockhounding. Yet they are often tied together in interests and the appreciation of understanding the natural world and all its complexities.

A plethora of reasons surround each geologic feature of the state, many of which are best explained properly by a geologist. There are many books that describe Montana's geology, one most highly recommended for the journey is *Roadside Geology of Montana,* by David Alt and Donald W. Hyndman. The book is an excellent companion to have on your rockhounding journey.

It is not the bottom of an ocean that people expect to find when walking across the high plateaus of Montana. Such items as sea shells, coral, and other marine life aren't what initially come to mind when we think of the land-locked northwestern states. Yet, there it is, a giant clam in the high desert, a shark's tooth on the side of a mountain! Marine fossils from the Idaho border to the Dakota line. Wait, and underneath the clams are terrestrial dinosaurs and plants, precious gems and mammoth bones. Even Lewis and Clark wondered why petrified skeletons of large "fish" they observed would be so far inland (one was a plesiosaur).

In a nutshell, Montana's geologic story is a tall tale. Scholars continue to understand more about the complex history of the state, but still there is much to learn. The oldest rocks in Montana, the basement rock, have been dated as far back as 3.2 billion years, and since that time, life-forms have come and gone, seas have filled and drained, mountains have been forced up, volcanoes have formed and erupted, glaciers have passed by, and the entire North American continent has changed shape.

There are some important geologic events in Montana to be familiar with. The basement rock may date from more than three billion years ago, but much of the rock you will come across is not this old. During Precambrian

Simplified Geologic Timetable

Era	Period	Epoch	Major Event
Cenozoic	Quaternary	Recent (10,000)	Ice Ages
	Tertiary	Pleistocene (2,000,000) Pliocene (11,000,000) Miocene (25,000,000) Oligocene (40,000,000) Eocene (60,000,000) Paleocene (70,000,000)	Age of Mammals
Mesozoic	Cretaceous (135,000,000) Jurassic (180,000,000) Triassic (225,000,000)		Age of Dinosaurs
Paleozoic	Permian (270,000,000) Pennsylvanian (305,000,000) Mississippian (350,000,000) Devonian (400,000,000) Silurian (440,000,000) Ordovician (500,000,000) Cambrian (600,000,000)		Age of Invertebrates
Precambrian	Proterozoic (2,500,000,000) Archeozoic (4,500,000,000)		Earliest Life (3,200,000,000)

(Numbers refer to years before present)

time (see geologic timetable), Montana was covered by an ancient sea. Sediments of this shallow sea produced the Belt series rock on top of the basement rock. For several hundred million years, these fine-grained deposits of sand and silt accumulated and eventually hardened into the slightly altered Belt series sandstones, and shales, that are visible throughout western Montana. Mudcracks, ripple marks, and other evidence of the shallow sea and even some primitive life such as algae can be viewed on these rocks today. These rocks deserve mention because you'll get to know them as you rockhound across Montana.

Transitioning to Cambrian time, shallow sea, freshwater, and terrestrial deposits continued to accumulate on top of the Belt. The first evidence of complex life appears and explodes. During the periods of the Paleozoic era that followed, thick beds of organic deposits in the east would eventually become some of the state's major gas and oil caches. Shallow seas continued to transgress and regress across Montana in to Mesozoic time, with a particularly notable inhabitant—the dinosaurs. The end of the Mesozoic era and the beginning of the Cenozoic era is marked by the abrupt disappearance of the dinosaurs. This dramatic change sixty-five million years ago is often referred to as the "K-T" boundary (an abbreviation of the boundary in time between the Cretaceous and the Tertiary). It is the subject of much controversy and study. There are only a few places in the world where the layer of sediments from this period of time can be viewed with the naked eye. Garfield County, Montana, is home to several of them.

As the Cenozoic moved on, massive vegetation deposits were buried in Montana, leading to the coal and lignite caches found in central and eastern Montana. In the west the Rocky Mountains began to form as the North American plate collided with the floor of the Pacific Ocean. Massive fault zones were created, batholiths emerged, and mountain-building events were in full throttle. Meanwhile, the east remains relatively quiet. A slight eastern uplift occurs as the Rocky Mountains formed, creating exposures of many of the fossiliferous layers we rockhound today. The Yellowstone hot spot erupted for the first time about two million years ago. The climate of Montana shifted from tropics to ice ages, mountains continued to form and erode, animals evolved, glaciers carved valleys, and eventually the landscape we know came to be. Now it is time to go rockhounding across Montana—a journey of three billion years.

These processes are in many ways what have created the minerals, fossils, and gems that we search for today. When at all possible in this book, a geologic

Western Montana is known for its mountains, valleys, and world-class trout fishing.

description has been included with each site to explain how specimens came to be and their relative age. Yet the most important aspect for the honorary geologist-in-training (which rockhounds are) is to ask questions about the rocks you find. "Where did this come from? How old is it? Why was it there?" "Oh dear, poor snail, how did he die?" "How does garnet form?" Once you've adapted to asking and answering questions, you can pull out a sapphire you mined personally and dazzle friends with your mineralogical expertise (field guide to gems and minerals required) or explain how your spectacular ammonite may have coexisted with the last dinosaur! These joys of discovery and provocative ways of learning are what will truly bring the pleasure of rockhounding to the soul. So prepare to quest for scientific knowledge. Soon the kids will be asking, "How old is a billion? Older than you?"

Montana's State Gems

Many states have an official gemstone; Montana has two. In 1969, agate and sapphire became the official state gemstones. These state stones are unique due to their abundance where just about anyone has the opportunity to collect them.

Montana Sapphire

Montana's sapphires weren't always a highly prized gem. During Montana's gold-rush era, sapphires were cursed for clogging sluice boxes and getting in the way of gold mining. Later as their value was learned, some gold operations, such as Montana's famous Yogo Gulch, were adapted to mine specifically for the sapphires.

Sapphires are a form of pure gem-quality corundum occurring in many colors, notably blue. In Yogo Gulch sapphires naturally occur in a rare deep cornflower blue. Montana's Yogo sapphire was the only North American gem to be included in England's "Crown of Jewels." The Yogo Mine is located on private land, off-limits to collecting, but there is barely a stop in Montana where a Yogo isn't featured on display or for sale. These unique gems are some of the most beautiful in the world; many sapphires found elsewhere are heat treated in an attempt to mimic the natural cobalt beauty of a Yogo.

For a detailed look at the history and geology of Montana sapphires, consult *Yogo: The Great American Sapphire,* by Stephen M. Voynick.

Outside of Yogo, there is still opportunity for the public to mine sapphires. This satisfying treasure hunt is highly recommended. There is nothing quite like wearing a sapphire you found yourself.

Montana Agate

There is so much to be told about agate. A chapter alone could be written on the various patterns and colors to be found in the magical stones, let alone the geology, history, and varieties. So it's a good thing that Montana has Tom Harmon (who owns the Agate Stop—Home of the Montana Agate Museum with his wife; see Appendix D), one of the original agate masters and author of the must-have-book *The River Runs North: A Story of Montana Moss Agate.* This book covers everything to know about agate and is full of some of the most stunning pictures of agate to be found.

Agate is a variety of quartz, more specifically a form of chalcedony that is very closely related to opal. In eastern Montana, agate is relatively common and was used by natives for thousands of years to fashion tools. Of course, they had impeccable taste, because the agate occurring in Montana is one of the more

Raw and cut Montana sapphires.

attractive and certainly a unique chalcedony variety to be found. The stone is popular due to its translucency, polishing capabilities, and wide variety of color and patterns. It can be found easily and has potential to be fashioned into jewelry, carvings, and other ornamental pieces.

"Montana Moss Agate" is the most sought-after variety of agate in the state. "Moss" refers to dendritic patterns that can occur in the translucent stones. Various images and beautiful landscapes can sometimes be seen in the patterning of the moss agate after it has been cut and polished. At any time of low water, agate can be commonly found throughout the gravels of the Yellowstone River. An eye must be developed to find a piece, as they often look like plain river rock due to their deceiving dull crusts. Hints of translucency can be seen in the sunlight—an instrumental tool to the search. A trip to Montana isn't complete without finding a beautiful moss agate. From pebble-size nodules to crystal-filled geodes and limb casts of trees, there is plenty to be found—perhaps the stunning piece for your next bola tie.

Montana Hodges and agate master Tom Harmon collect agate on the Yellowstone River in late summer.

Montana agate is often collected in the gravels of the Yellowstone, but it was formed long before the time of the river. Most Montana moss agate is believed to have been created about sixty million years ago during volcanic activity in what is today the Yellowstone Park area. Agate can form as volcanic flows cool; in some way or another, water rich in silica would flow through lava and fill cavities. Perhaps the liquid silica would fill just an air pocket, or maybe a cast of a buried tree limb, and eventually harden into agate.

Rockhounding

Rockhounding is a hobby for a special person. To some, the "hunting" of rocks can seem quite bizarre. After all, if it was just rocks someone is after, they're everywhere. Rockhounds collect many different kinds of rocks for many different reasons; some collect fossils, or rocks, or minerals, and some don't keep anything at all. To many, rockhounding can be much more than finding rocks. It can take you to locations beyond the standard roadside attractions and allow you to explore areas of America that many tourists never see. Rockhounding pushes you through towns where outsiders seldom come, on roads without pavement, past cattle trails, to ghost towns, to abandoned quarries, to forgotten mining districts, and to mountaintops. Perhaps rockhounds are the last of the American explorers.

The best part is that anyone can be a rockhound. Whatever your flavors from ammonites to micromounts, whether you're looking to collect for an hour or a month, stay close to the car, or prospect the Rockies, there is a site in Montana for you. And maybe you'll learn something new or bring home a gigantic sapphire. Rockhounding is a great family sport. You don't need television, junk food, soda, or a scoreboard (although if you bring all of these along the group may last longer). I have never run across an educational pastime as entertaining as looking for rocks. So bring the kids. The more kids the better; they're closer to the ground—they find better stuff! It's treasure hunting with an opportunity to learn. Here is your chance to make science fun and tell the kids that they *can* touch things.

Collecting Regulations and Etiquette

Public lands make up much of Montana's landscape, creating millions of acres open to people for exploration. Rockhounding is allowed on several forms of public land. In general, rockhounding is allowed on Bureau of Land Management (BLM) and forest service land, with restrictions. Rules and regulations vary by location and could change at any time, so always inquire before collecting.

It may seem like a pain, but visiting every BLM office and ranger station is all part of the experience. Oftentimes there are multiple benefits to stopping at the office. Aside from checking land status, the staff can be helpful. They may give you great advice on where to rockhound, camp, hike, or lunch. An insider's tip could become the highlight of the trip.

There are some basic regulations to be familiar with while collecting on public land. Collecting or disturbing vertebrate fossils on government property is not allowed without a proper permit. This rule is strictly enforced. Also, all findings are subject to the Federal Antiquities Act of 1906, and anything of historical value should not be removed. Anything of archaeological nature is protected under the Native American Graves Protection Act and should not be removed. This means that an arrowhead made of agate is not legal to collect. It may have been agate first, but it is an artifact now. The good news is that any rock, gem, mineral, or fossil that you do find while abiding by the rules is yours to keep (like that two-ounce nugget!).

USDA Forest Service Land

Rockhounding is permitted on national forest land, although rules and regulations vary by location. These guidelines are ever-changing, so always contact the local forest service office for an updated map (see Appendix C) of the area you are collecting in and inquire about the land status.

In general, rockhounding and gold panning can be done without a permit, fee, or special permission. Any kind of mechanical equipment and blasting are prohibited. All activity must be done without significantly disturbing the environment. A good example of what is against regulations would be digging excessively into tree roots for crystals. Many trees die from this activity, creating safety hazards and an ugly public forest.

As on all federal land, collecting vertebrate fossils is not allowed. Material that is allowed to be collected must be taken for personal use with no commercial purpose. Size regulations vary by location. Exact collecting limits are not generally specified, but quantities are required to stay small. A few rocks are fine, but enough to make a new pathway through your rose garden probably requires a permit. A good example of a variation to the standard rule is the Gallatin Petrified Forest. A permit is needed to collect petrified wood in the boundaries of Gallatin Petrified Forest, and specimens collected are limited to 20 cubic inches per person per year.

Bureau of Land Management Land

The same rules apply to BLM land, although in this case, there are specified limitations on the amount of petrified wood. No more than 25 pounds per day, at a maximum of 250 pounds per year, can be taken per person from BLM land. This is still a generous offer, but remember that all collecting must be done for personal hobby use, just as on national forest land.

Other Federal Land

Collecting within national parks, monuments, wildlife refuges, and military land is not allowed without a permit for a qualified party. These rules are highly enforced, especially in areas with great geologic significance.

State Land

State land may be open or closed to the public. Always inquire for status before entering state land. When state land is leased, it may function similarly to private property, and trespassers can be prosecuted if they do not obtain permission from the lessee. If it is open to the public, rockhounding is still restricted and collecting of anything on the land probably requires a permit.

Tribal Land

Tribal land also functions similarly to private land. In order to enter or collect, you must obtain permission through the Office of the Superintendent of the Bureau of Indian Affairs associated with the reservation in question, and if approved, the landowner of the particular location.

Private Land

Private land is simply private. Unless permission is obtained from the landowner, private land is not accessible, much less available, for collecting. Never assume that because there aren't signs or fences that the land is open for collecting. In some areas of Montana, you can drive for miles through private land and never see a fence (you may see the landowner's cows in the road, however). If permission is obtained from the landowner to collect on the property, anything you find belongs to the landowner unless the landowner relinquishes ownership.

Mining Claims

Mining claims function similarly to private land, even when they are located on public land. Often the public is allowed to pass through the physical property but, unless permission is obtained from the claim owner, disturbing or removing material from the claim is not allowed. Respect others claims and be on the lookout for their postings. Mining claims are generally posted on small poles or tree trunks. To avoid trespassing, it is always best to stay off and away from anything you suspect is claimed.

How to Use This Book

Maps

The regional maps found in this book have latitude and longitude lines to help find the general location of the rockhounding sites and to assist with GPS usage. The importance of having a detailed map cannot be overemphasized. In many parts of rugged Montana, an atlas isn't sufficient. A forest service, BLM, or other kind of large-scale map of backcountry roads is needed for many of the sites. A wrong turn on an unlabeled forest road could lead to hours of despair. Maps can be ordered in advance (see office locations in Appendix C) or purchased from various offices during their business hours. A reference to the *DeLorme: Montana Atlas & Gazetteer* is included at the end of the "Finding the site" category in every site write-up for convenience.

Using the Global Positioning System

New to this book is the Global Positioning System (GPS) coordinates at each site. A GPS is a wonderful tool to have, and many new vehicles come equipped with one. The system works off satellites and gives the degree coordinates of latitude and longitude of its location with average accuracy of around 20 feet. A fraction of a degree can be read as minutes and seconds or as a decimal. This book lists the coordinates in the decimal system, so set your GPS accordingly. The inexpensive handheld GPS is popular with hikers and is highly recommended for any trip into the wilderness.

When possible, the coordinates listed for the sites in this book are at the most productive location. If the site involves you parking and exploring, the reading may be from the parking area. The coordinates listed are intended to get you within a few hundred feet of the area. Once you arrive you then must use your rockhound skills and luck to find the treasure. Also remember that exposed rocks and productivity change consistently with weathering and erosion, sometimes drastically, so exact locations are variable. Specific GPS coordinates are not provided for sites with mineral deposits or material that are spread out over wide areas or are sporadic.

Season

The recommended time to visit each site is often described by season. General terms are used instead of specific months because of the unpredictable of snowmelts, early winters, etc. Spring refers to the time after the snow melts enough to drive on the roads, which could occur anytime from April through June depending on the year and elevation. Summer is when snow is no longer falling. Fall is when the temperature reaches freezing but the roads are still accessible. It is important to inquire locally about road conditions and check local weather forecasts before trekking out into the wilderness, especially at high elevations.

Weather

Montana's weather can be unpredictable and extreme. Thunderstorms are common during the summer months, so be prepared for the safety hazards associated with this. Each site was visited during good road conditions, and consequently, the descriptions of road conditions apply to nice weather. Recent storms, rain, snow, debris, and many other things can affect accessibility. A road that may be listed as suitable for any vehicle could become a high-clearance four-wheel-drive jeep trail when muddy. Use your best judgment, or call it a day when the dark clouds roll in.

Where to Stay

A Montana map can be deceiving. Many names on the map that appear to be towns turn out to be fields (or abandoned train stops) when you get there. This is a common problem with visitors to the more remote areas of the state. Bold print doesn't necessarily mean anyone will be there, much less a hotel. It is important to research your stops and secure a place to stay well in advance, whether it's a campsite, RV space, or motel, all of which can be booked solid during summer months.

Safety

Safety in rockhounding involves, as it does in most outdoor hobbies, exercising common sense and being prepared. The weather in Montana can be unpredictable, sunny weather can quickly switch to flash floods and lightning. Montana is also home to rattlesnakes, grizzly bears, huge mosquitoes, deer flies, gnats, and many other dangerous and pesky creatures. Always make your presence known in bear country and watch where you put your feet, especially in areas frequented by rattlesnakes. Take a can of bug repellent with you wherever you go.

Bring the proper equipment and protection to each site: safety goggles, gloves, and a first-aid kit. Be prepared for getting lost or stranded. There is nothing more embarrassing than being stuck in a pile of snow on isolated Sweetwater Road outside of Dillon, Montana, at 10 o'clock at night without a flashlight or shovel. Basic supplies are necessary, and cell phone reception cannot be relied upon, especially outside of big cities.

One of the biggest safety hazards that rockhounds face are abandoned mines. The motto is "Stay out—Stay Alive" and it couldn't be more aptly summed up. Mines are extremely dangerous and should be avoided. Many minerals occur in the tailings piles around mines, and it would be a shame for the poor judgment of those who enter these dangerous mines to risk making collecting in the nearby areas restricted. Never go inside an old mine. Everything good has already been removed. If a site has known shafts, that fact will be mentioned, and children should skip these areas.

Most public land in Montana that is open to rockhounds is also open for hunting during certain times of the year. Always inquire about hunting season, which is generally in winter when the snow is too heavy to rockhound in remote areas anyway. If you do visit a site during hunting season, it would be wise to wear orange safety vests and other bright colors while in wooded areas.

Ten Vistas for Your Montana Journey

Glacier National Park

You can't collect here, but people from across the world come to view the mountains and glaciated peaks of northwestern Montana's Glacier National Park. The park preserves more than a million acres, including some of the most breathtaking landscape. Lush, green valleys carved by the ice ages sit below towering mountains. In summer hundreds of species of wildflowers bloom in these picturesque lowlands. Dozens of lakes line the mountains, and approximately fifty glaciers dot the mountains. Some of the glaciers are visible from the road, and others can be seen through more than 700 miles of hiking trails. You can also find some of the oldest fossils in Montana, Precambrian stromatolites (algae) fossils, within the trails and road cuts. The park is home to almost every kind of large mammal native to the United States and more than 200 bird species. While there don't forget to travel on the "Going-to-the-Sun Road," a scenic route through the park that crosses some of America's best photographic opportunities. The park is generally open from late May to mid-October.

Getting there: The park is located in northwestern Montana on the U.S.–Canadian border. The park and its headquarters can be accessed from U.S. Highway 2 in the west; there is also an entrance from US 2 on the eastern side of the park.

Yellowstone National Park

Although most of Yellowstone National Park is actually located in Wyoming, three of its five entrances are in Montana, and the park represents much of the state's tourism. Yellowstone is visited by people from around the world, and it's home to one of this country's largest wildlife sanctuaries. The park is also one of the world's largest active volcanoes. A rockhound can really appreciate this unique chance to view the magnificence of the volcanic process in one of the grandest thermal basins. Spurting geysers, terraced hot springs, and even boiling mud can be viewed easily. Wildly colored thermal pools and "screaming" vents of heated gases endure gloriously for generations of photographers. Mammoth Hot Springs, Old Faithful, and the Grand Canyon of the Yellowstone are some of the park's more famous destinations. Yellowstone is also home to many species of wildlife, including the once-endangered gray wolf, and the park is noted for being one of the most successful wildlife preserves in the world. More than 2,000 bison roam the area, and they often leisurely stroll along the road. Beyond the pavement, more than 1,200 miles of trails cross the park's many wonders. The park is open for automobile travel from May through September. Limited access can be gained by other modes of transportation or limited automobile access during the off-season.

Getting there: The park is located in the southwest corner of Montana and has five entrances. The headquarters are located in Montana at the north entrance through Gardiner on U.S. Highway 89. The northeast entrance to the park is through Cooke City, and the west entrance to the park is located in West Yellowstone off U.S. Highway 191. The park can also be reached from the towns of Jackson and Cody in Wyoming.

Beartooth Pass and the Beartooth Mountains
The Beartooth Highway (U.S. Highway 212) east of Cooke City is a scenic 64-mile stretch of road that connects Red Lodge with Cooke City and the northeast gate to Yellowstone National Park. The Beartooth Pass crests at more than 10,000 feet, an elevation well above the tree line, and this is one of the most beautiful drives in the state. Here, high mountain country and the occasional summer snow create wonderful photographic opportunities. The Native Americans of the area called the climb the "trail above eagles."

The Beartooth Mountains surrounding Cooke City are noted for their unique geological features. Active glaciers of any significant size no longer exist here, but a few small remnants are present. Of these, Grasshopper Glacier is perhaps the most interesting. Apparently, at the time glaciers were active in these mountains, swarms of locusts became trapped in the heavy snows and were frozen. Eventually they were buried by additional snowfalls and were incorporated into the glacial mass. Many layers of these creatures can be viewed in this glacier, particularly during the late summer months when the ice is better exposed. It is a hike of considerable distance to the glacier though, since the wilderness designation of the area does not permit motorized travel, but it makes a most enjoyable outing.

Getting there: US 212 connects Red Lodge with Cook City and can be reached either heading south out of Red Lodge or from Cooke City just above the northeast entrance to Yellowstone National Park.

The Butte Area
The Butte area is probably the most exciting place in Montana for attractions that rockhounds can really appreciate. The World Museum of Mining and Hell

Roaring Gulch just outside the town limits of Butte is built on the grounds of the Orphan Girl mine. The museum has many displays on mining history and a reconstructed 1800s mining town with more than fifty buildings known as "Hell Roaring Gulch." Yet, the "gold mine" of Butte's special attractions is the Mineral Museum on the grounds of Montana Tech University. Along with geologic history and information, the museum has 1,500-plus minerals on display, including a fluorescent minerals room, 400-pound quartz crystals, and what may be the largest gold nugget ever found in Montana. Also while in the area, don't forget to stop by the free Berkeley Pit Viewing Stand to view the consequences of excessive rockhounding. The giant crater stands as testimony to what was one of the world's largest truck-operated mines. Between the 1950s and 1980s, the Berkeley Pit produced more than $2 billion worth of gold, silver, copper, and zinc.

Getting there: The World Museum of Mining and Hell Roaring Gulch is located at 155 Museum Way in Butte. The Mineral Museum is located on the grounds of Montana Tech inside the Museum Building at 1300 West Park Street. The Berkeley Pit Viewing Stand is located on the eastern end of Park Street. All attractions are open year-round; inquire about exact times.

Bighorn Canyon

Beautiful Bighorn Canyon and the Pryor Mountains west of it are located in the high desert on the Montana–Wyoming border. The Bighorn Canyon National Recreation Area is a wonderful stop for water sports, walleye fishing, and handsome scenery. It was named for its deep canyon walls along 71-mile-long Bighorn Lake. There is much wildlife in the area, from world-class fishing in the Bighorn River to wild horse viewing in the Pryor Mountain horse range just outside it. Keep an eye out for bald eagles, one of hundreds of bird species to be seen in the canyon. Rockhounding is not allowed within the canyon, so include a stop at the Pryor Mountains (see site #89) to collect some fossils.

Getting there: Bighorn Canyon, although in Montana, is best accessed through Wyoming Highway 37 north of Lovell, Wyoming. The canyon can also be reached from Montana by driving about 40 miles south of Hardin on Montana Highway 313.

Fort Peck Dam Interpretive Center and Museum

Fort Peck Lake is home to one of the largest earth-filled dams in the world. New to this area is the impressive Fort Peck Dam Interpretive Center and Museum, located just below the dam. The center works together with Fort Peck Paleontology Inc. to host one of the most thrilling dinosaur displays in the state. Upon entrance to the enormous building, a life-size fleshed-out model of the Tyrannosaurus rex known as Peck's Rex greets visitors. Behind the T-rex, gigantic windows overlook the sublime lake. The museum holds many other paleontology displays along with exhibits on Montana history and wildlife and Fort Peck.

Getting there: The museum is located in Fort Peck just south of the dam, between the powerhouses and campgrounds.

Museum of the Rockies

The Museum of the Rockies is located on the grounds of Montana State University (MSU) in Bozeman. The museum has displays on many aspects of northern Rockies history, with a special emphasis on geology and detailed geologic exhibits with beautiful fossil specimens. The paleontology displays include the skulls of some of Montana's most famous species of dinosaur. Also in the museum are a planetarium and many exhibits on Montana history, including a large outdoor Lewis and Clark display. The Museum of the Rockies is open year-round; inquire about times and holidays.

Getting there: The museum is located in Bozeman on the MSU campus at South 7th Avenue and West Kagy Boulevard.

Gallatin Petrified Forest

It is a good hike of several miles round-trip to go there, but the Gallatin Petrified Forest north of Gardiner is worthy of the climb (see site #56). Here an ancient forest that was once covered in ash, which turned much of the original tree material into petrified wood, has slowly been exposed through thousands of years of erosion. Magnificent tree trunks can be seen, and *with a permit* small amounts of material can be legally collected. Some of the material is of exceptional quality, but the highlight is by far the journey into the forest of the past via a beautiful mountain valley.

Getting there: The trailhead for Gallatin Petrified Forest is located 16 miles north of Gardiner off U.S. Highway 89. Once you pass the small town of Miner, head 10 miles west on Tom Miner Basin Road.

Sun River Canyon

One of Montana's most secluded gems, the isolated Sun River Canyon is an easy escape from the ordinary. The canyon isn't off a road leading toward any popular destination, and most likely it is out of your way, but the atmosphere alone is worth the journey. The drive from Augusta to the canyon is stunning. Rolling high prairies and some of the biggest and clearest sky of the state lead to a dramatic shift of the surprising mountain range ahead. Within the range the Sun River and amazing limestone cliffs of the canyon await. Large Gibson Reservoir above the dam has beautiful sandy beaches and excellent fishing opportunities. Bring binoculars to observe the overwhelming amount of mountain wildlife. Hiking, camping, fishing, and rockhounding are all excellent at this location.

Getting there: The canyon is located 15 miles north of the small town of Augusta (west of Great Falls), on Sun River Road.

Makoshika State Park

Makoshika is Montana's biggest state park and by far the most interesting one. The name is from the Sioux language for bad land. The park preserves more than 11,000 acres of eroded earth deemed badlands. Being anywhere in the park is like being in another world. Badlands are soft sandstones and shales formations sculpted by weather. The highly eroded high desert and plains formations look much like a landscape of a science fiction novel. Giant sandstone boulders rest upon thin columns of earth, waiting to fall. Wash beds, alluvial fans, and crumbling hills of sands are interrupted only by the occasional coulee, spring, or desert vegetation. At sunset the sage-tinged placid landscape fades to colors of pink, orange, and red.

Many dinosaur finds, including several of significant scientific value, have been discovered in the Cretaceous Hell Creek formation in the park. A small visitor center and museum located just within the gate to the park house displays on the geology and fossils of Makoshika. The park is open year-round, weather permitting.

Getting there: The entrance to the park is located 1 mile south of Glendive at the end of Snyder Avenue.

Map Legend

Administrative Boundaries

──── State/Province Boundary

──── County Boundary

─··─ Indian Reservation

National Park/Forest

Wilderness Area

Transportation

═══ Multilane Divided,
Controlled Access

─── Two Lane, Limited Access

─── Two Lane Paved

─── Secondary Two Lane Paved

─── Unpaved Road

------ Trail

🔟 Interstate Highway

🔟 U.S. Highway

🔟 State Highway

228 Secondary State Highway,
Local Road, or Forest Road

Hydrology

⬭ Reservoir or Lake

─── River

─── Creek

─··─ Intermittent

Population

⊛ State Capital

⊙ City

○ Village

Physiography

)(Pass

▲ Peak

∩ Cave

······ Continental Divide

Symbols

❷ Site Number

🅿 Parking

△ Campground

▲ Backcountry
Campground

⚑ State Park with
Camping

❓ Information

🛈 Ranger Station

⌂ Rest Area

🚶 Battlefield

🏃 Rockhounding Site

▪ Point of Interest

👁 Scenic View

⚒ Mine/Prospect

⛷ Ski Area

≍ Bridge

⌣ Dam

✈ Airport

🐾 Wildlife Area

Whitefish Mountains

See map on page 24
Land type: Mountains, road cut.
GPS: N. 48.93865 / W. 114.54911
Best season: Summer.
Land manager: USDA Forest Service, Flathead National Forest.
Material: Marine fossils.
Tools: Rock hammer.
Vehicle: Any.
Accommodations: Camping and RV parking in Flathead National Forest; camping, RV parking, and motels in West Glacier.
Special attractions: Glacier National Park.

A fossilized horn coral found in a loose piece of limestone.

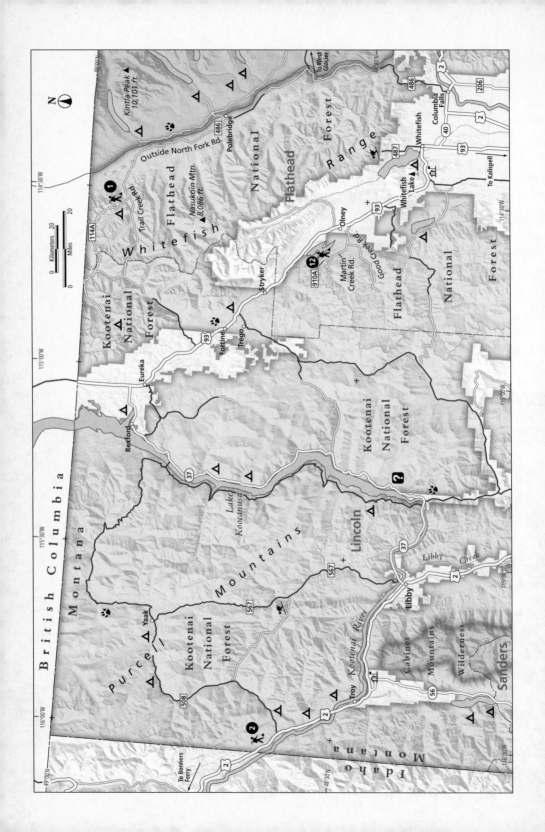

Finding the site: This site is a tough one to find; a good forest map will be needed. The location is at a high elevation not far from the Canadian border; local inquiry should be made if plans are to visit in either spring or fall.

From Polebridge drive about 15 miles north on Outside North Fork Road (Montana Highway 486) and turn left onto Trail Creek Road. Drive 5 miles and turn right onto Thoma Creek Road (Forest Road 114A). Drive 1.2 miles up Thoma Creek Road to the site. It is difficult to spot the limestone outcrops on the right side of the road because the mountainside is covered in lush vegetation. Subsequently, mileage and exploration is very important. Also keep an eye out for pieces of loose limestone near the edge of the road that might contain small fossils. *DeLorme: Montana Atlas & Gazetteer:* Page 83 A5.

Rockhounding

The geology of far northwestern Montana consists almost entirely of Belt rock and occasional diabase sills. In the area north of Polebridge, high in the Whitefish Mountains, within walking distance from the border with British Columbia, there are unique deposits of Cambrian limestone. Some of the limestone contains horn coral and brachiopod fossils. The coral is quite large, more than an inch in diameter, and is fairly well preserved. Yet, diligent searching is required to locate both the site itself and fossiliferous rock. Splitting the hard limestone to expose or remove an intact fossil is very difficult as well. The best fossil hunting is likely done by investigating loose pieces of limestone beneath or near the road cuts.

Troy

See map on page 24
Land type: Mountains, road cut.
GPS: N. 48.61547 / W. 116.02937
Best season: Spring through fall.
Land manager: U.S. Highway, maintained by Montana DOT.
Material: Precambrian mudcracks, ripple marks.
Tools: None.
Vehicle: Any.
Accommodations: Camping, RV parking, and motels in Troy, Montana, and Bonners Ferry, Idaho.
Special attractions: Kootenai Falls, located between Libby and Troy on U.S. Highway 2, is one of the largest undammed falls in the northern Rockies.
Finding the site: From Troy drive 12.5 miles west on US 2 toward Idaho. Several small road cuts of vertically turned mudstone layers are exposed along the right side of the road. Park your vehicle safely off the highway on one of the small roads in the area and walk to the site. Use caution around highway traffic. *DeLorme: Montana Atlas & Gazetteer:* Page 81 B4.

Rockhounding

More than a billion years ago, before humans walked the earth, before dinosaurs existed, before bugs, before clams, before worms, before trees, and before plants, the Belt rocks were being formed. The slightly metamorphosed sedimentary rock layers known as the "Belt series" developed from 600 million years of sandy and muddy deposits. Fine surface details in the layers of theses rocks include mudcracks and ripple marks—but you will find no fossils. When the Belt formations were deposited, the landscape contained nothing that left a fossil or a track. By looking at the ancient mudflat remains, you can see the evidence for a very different earth than we know today. Even the atmosphere of the earth was unlike that of today. The climate was scalding hot and filled with acid rain and little wind. These ancient Precambrian sedimentary samples are not as desirable as fossils, but every rockhound should have one for their collection, just to have a good piece to take out sometimes and reflect on the changes of our planet.

At this site the layers of green and red mudstone break apart very easily with the assistance of a rock hammer. Some of the best samples collected were found at the bottom of the road cut, naturally split by weathering.

Libby Creek

See map on page 28

Land type: Mountains, creek.

GPS: N. 48.11034 / W. 115.54579

Best season: Late spring through summer.

Land manager: USDA Forest Service, Kootenai National Forest.

Material: Gold, garnet.

Tools: Gold pan, trowel, fleck flask.

Vehicle: Any.

Accommodations: Camping and RV parking at Howard Lake; camping, RV parking, and motels in Libby.

Special attractions: The Heritage Museum in Libby has exhibits on mining, forestry, and local history.

Finding the site: From Libby drive 13 miles south on U.S. Highway 2. Turn right onto Libby Creek Road (Forest Road 231) and drive 10 miles to the main parking area. Large signs posted near the primitive restrooms display a map of the area open for recreational panning and, more importantly, the area that is private and off-limits. The creek is located on one side of the parking area. A very informative brochure along with a map is available at the Libby Chamber of Commerce and is highly recommended. The staff can also tell you about different gold-prospecting stores in the area if equipment needs to be purchased.

For a brochure on Libby Creek Recreational Panning Area, call or write to Libby Area Chamber of Commerce, P.O. Box 704, 905 West 9th Street, Libby, MT 59923; (406) 293–4167. *DeLorme: Montana Atlas & Gazetteer:* Page 80 A3.

Rockhounding

Mining began in the Libby Area with placer deposits worked for gold in the early 1880s. The oldest lode mine was developed about 1887, and since that time mining has been sporadic and apparently not extremely profitable. Most of the mines were located about 20 miles south of Libby in the tributary valleys along the east slopes of the Cabinet Mountains and south of Troy (to the west of Libby). Unfortunately, these mines and tailings offer little opportunity to the rockhound as most are private. Despite the lack of mine tailings, the Forest Service has set aside this wonderful site for the public to pan out their own gold. Unlike the hard-rock gold mines, the Libby Creek Recreational Gold

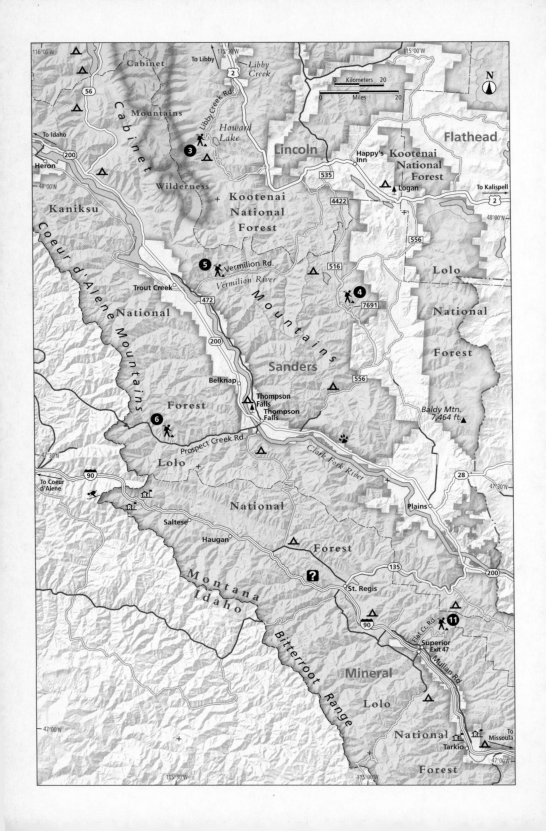

Panning Area is rich in placer gold deposits. Since the Pleistocene era, and until about 10,000 years ago, glaciers that once covered the area eroded away the edges of mountains and carried rock debris into what later became the Libby Creek drainage area. Gold, along with other heavy minerals, became concentrated as the material was worked through the flowing waters, leaving behind the heavy metals.

Plenty of gold can still be found along Libby Creek today. The site is open to the public and free of charge. Any gold you find is yours to keep. Prospecting is limited to panning and hand tools. No motorized or mechanized mining is allowed. While panning for the ancient gold in the cool water, examine the Precambrian rock, now smooth from the long journey. You can almost imagine what the landscape must have been like when the ancient Belt rocks were being carved by glaciers.

If you catch a hint of "gold fever" while on Libby Creek, keep in mind that the Northwest Montana Gold Prospectors club also owns a claim on Libby Creek and, for a small fee (during my visit it was $20 per year per family), their extensive claims are also accessible. For information contact Northwest Montana Gold Prospectors (NWMGP), P.O. Box 3242, Columbia Falls, MT 59912; (406) 892–3722.

The mining that perhaps made Libby most famous was the extensive deposits of hydrous biotite mica, called vermiculite, that were mined several miles northeast of Libby from the 1920s to 1990. When active, the mine supplied 80 percent of the vermiculite used worldwide. Vermiculite is a material that expands when heated. It has many uses; its primary use was as an insulation material. Almost ten years after the mine was out of operation, the Environmental Protection Agency determined that the vermiculate was a host of airborne cancer-causing asbestos and responsible for asbestos-related disease and fatalities. Today the Environmental Protection Agency oversees the Libby Superfund cleanup site.

Fishtrap Creek

See map on page 28
Land type: Mountains, road cut.
GPS: N. 47.82308 / W. 115.14895
Best season: Summer.
Land manager: USDA Forest Service, Kootenai National Forest.
Material: Fossils.
Tools: Rock hammer.
Vehicle: High clearance.
Accommodations: Camping and RV parking in Kootenai National Forest; camping, RV parking, and motels in Libby.
Special attractions: The Heritage Museum in Libby has exhibits on mining, forestry, and local history.
Finding the site: Some of these forest roads are difficult to locate so a forest map is recommended. From Libby drive southeast on U.S. Highway 2 and turn right onto McKillop Road (Forest Road 535) toward the Fishtrap Creek (not lake) campground. The road will change numbers from FR 535 to Forest Road 4422 and then to Forest Road 516; the important thing to remember is to follow the signs to Fishtrap Creek (not lake) campground. Drive a total of 22.5 miles, turn left onto Shale Ridge Road (FR 7691), and continue 0.7 mile to a hill on the right side of the road with very thin gray shale layers eroding. Look carefully—the shale is so thin and easily weathered that it looks like dirt. *DeLorme: Montana Atlas & Gazetteer:* Page 66 B2.

Rockhounding

The paper-thin layers of Cambrian shale contain fossilized casts of trilobites, a marine arthropod characterized by a beetle-looking three-lobed ovoid outer skeleton. Trilobites are from the Lower Cambrian (beginning of the Cambrian time) and their existence stretched to the Permian period. Trilobites existed for about 345 million years and in the end more than 4,000 species evolved, but with very little variation. The Libby-area trilobite fossils vary from about ¼ to ⅜ of an inch in size and less commonly up to 1½ inches in size. Most trilobite fossils are actually a fossil of the shed skeletons and furthermore a cast, or internal mold, of the shed skeleton. On rare occasion the complete trilobite can be found, and on an even rarer occasion a fossil of the underbelly of the trilobite is found. Nevertheless, these little critters are excellent additions to fossil collections. Brachiopods are less common than the trilobites but can be found at this site as well.

The shale emerging from the road cuts is difficult to spot. Like many brittle shales, it erodes at a rate so extreme that it appears as dirt. Layers containing fossils can be exposed by brushing away shale debris and exposing hard layers of shale. The shale is still extremely brittle and, during my visit in July, it was also very wet. The layers have to be carefully split to expose trilobites and brachiopods and, due to their delicacy, diligent searching is required to collect a complete sample. Bring packing material and a secure container to transport delicate specimens.

Vermilion Road

See map on page 28
Land type: Mountains, river.
GPS: N. 47.860171 / W. 115.45975
Best season: Summer.
Land manager: USDA Forest Service, Kaniksu National Forest.
Material: Gold.
Tools: Gold pan, shovel, bucket, fleck flask.
Vehicle: Any.
Accommodations: Camping and RV parking in Kaniksu National Forest; camping, RV parking, and motels in Trout Creek.
Special attractions: None.
Finding the site: A suggested location is reached by driving northwest from Trout Creek on Montana Highway 200 for about 1.5 miles and turning right onto an unmarked road (Montana Highway 472). Drive 4 miles and turn left onto Vermilion Road. Drive 5 more miles and look for a nice place to dig. Particular luck was had in Lyons Gulch. Look for places where others have been digging and a nice quiet shore to pan out your material. *DeLorme: Montana Atlas & Gazetteer:* Page 80 C3.

Rockhounding

A giant granite intrusion in Precambrian Belt rock is responsible for the occurrence of the gold in the area around the Vermilion River. It wasn't until 1887 that a group of Montana prospectors actually found the deposits of gold. The area never produced enough profit to sustain mining operations, and plenty of material is still there today for the recreational panner.

The deposit area begins just a few miles up Vermilion Road. The easiest way to spot a good area is to look for where others have been digging. At road cuts, at tree roots, at soil that appears to be from an older stream flow, and even in the riverbed itself under big rocks or at bends are the best places to dig and fill a bucket.

Cox Gulch

See map on page 28
Land type: Mountains.
GPS: N. 47. 54753 / W. 115.59152
Best season: Spring through fall.
Land manager: Private, United States Antimony Corporation.
Material: Stibnite, pyrrohite, galena, and sphalerite.
Tools: Rock hammer.
Vehicle: Any.
Accommodations: Camping and RV parking nearby in Lolo National Forest; camping, RV parking, and motels in Thompson Falls.
Special attractions: The restaurant, lounge, and lobby of the Rimrock Motel in Thompson Falls has a mining-theme decor, with several glass cases of minerals on display and available for purchase.
Finding the site: From Thompson Falls drive west on Montana Highway 200 and turn left onto Prospect Creek Road, just west of the Clark Fork River on the outskirts of town. Drive 13.5 miles and turn right at Cox Gulch and drive to the parking area. Inquire at the United States Antimony Corporation (USAC) office located in the parking area for access. It is recommended that you write in advance for permission to check tailings, or call ahead to ensure your arrival during the office hours. For more information contact the United States Antimony Corporation, P.O. Box 643, 1250 Prospect Creek Road, Thompson Falls, MT 59873; (406) 827–3523. *DeLorme: Montana Atlas & Gazetteer:* Page 80 D3.

Rockhounding

Though numerous mining districts can be found within Sanders County, those in the immediate vicinity of Thompson Falls are currently the most important. The earliest mining probably began

Stibnite in quartz.

The U.S. Antimony Corporation processing plant and office; inquire before collecting.

around here in the 1880s, during the Coeur d'Alene gold rush. Little is known concerning production then, but it is certain that fair quantities of gold were recovered from placer deposits. Today the Prospect Creek District southwest of Thompson Falls is of particular importance because of the presence of antimony-bearing quartz veins within the Precambrian rocks. Mines are being developed by the U.S. Antimony Corporation and are conveniently easy to access through the USAC office. Permission to dig through the tailings must be requested either in person or by mail, and at the time of my visit it was nicely granted. Someone in the office will recommend directions to the tailing piles to ensure no disturbance on active areas. At the time of my visit a dump pile of low-grade ore produced nice samples of stibnite, the principle ore of antimony, and stibnite crystals. The stibnite has a bright metallic silver luster, and the prismatic orthorhombic crystals are small but worth the search. With some luck, micromount quality of other minerals such as pyrrohite, galena, and sphalerite may be found.

Bynum/Choteau

See map on page 36
Land type: High hills.
GPS: See contact information below.
Best season: Generally operates May through September.
Land manager: Private, Timescale Adventures.
Material: Fossils.
Tools: None.
Vehicle: Any.
Accommodations: Camping, RV parking, and motels in Choteau.
Special attractions: Old Trail Museum in Choteau has exhibits on local history and paleontology. The museum also offers hands-on dinosaur digs and field courses.
Finding the site: Digs are generally located in the Bynum/Choteau area of Montana along the Rocky Mountain front. The organization is centered out of the Two Medicine Dinosaur Center in Bynum. Contact Timescale Adventures for updated information. *DeLorme: Montana Atlas & Gazetteer:* Page 69 B8.

Rockhounding

Timescale Adventures is a nonprofit organization based out of the Two Medicine Dinosaur Center in Bynum, located in the heart of the Two Medicine Formation. The center has extensive displays on paleontology and is home to a skeletal model of the world's longest dinosaur. Paleontology field trips into the Two Medicine Formation of the Choteau area with Timescale Adventures can range anywhere from a couple of hours to a ten-day course. A minimum of a daylong program is required to participate in an excavation. The courses are designed as "hands-on," and the staff incorporates an overview of geologic history, fossil identification, excavation, and preservation. Longer programs cover all aspects of field paleontology. Group sizes are generally limited, and participation by children may be restricted to those fourteen years of age and older. Reservations are recommended well in advance. For more information contact Timescale Adventures, P.O. Box 786, 120 Second Avenue South, Bynum, MT 59419; (406) 469–2211, (800) 238–6873, www.tmdinosaur.org.

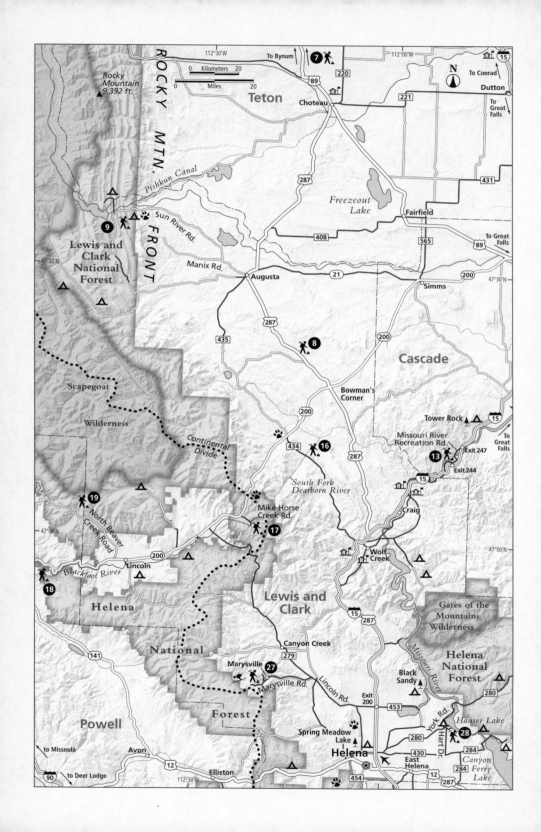

Augusta

See map on page 36

Land type: Low mountains, road cut.

GPS: N. 47.38915 / W. 112.31542

Best season: Late spring through fall.

Land manager: Montana DOT.

Material: Fossils.

Tools: Rock hammer.

Vehicle: Any.

Accommodations: Camping and RV parking in Lewis and Clark National Forest by Augusta; camping, RV parking, and motels in Great Falls.

Special attractions: Sun River Canyon, located 15 miles north of Augusta on Sun River Road.

Finding the site: From Bowman's Corner at the junction of Montana Highway 200 and U.S. Highway 287, drive north on US 287 for 10.2 miles to a road cut on the right. The most fossiliferous layers are located on the closest side of the road cut to Bowman's Corner, and there is no need to pass the bend. It is best to rockhound on this side of the cut; otherwise, a blind spot is created for oncoming traffic. Exercise caution when anywhere around speeding traffic. This site is not recommended for children. *DeLorme: Montana Atlas & Gazetteer:* Page 55 A8.

Rockhounding

In the area near Augusta, a thick concentrated layer of solid Cretaceous oyster fossils in sandstone can be found. The oysters still retain their mother-of-pearl shell, but they are almost impossible to break from the road cut. The best luck collecting good specimens is found by searching for fossils in loose pieces that have already fallen loose from the parent rock.

Sun River

See map on page 36
Land type: Mountains, road cut.
GPS: N. 47.61869 / W. 112.70161
Best season: Late spring through fall.
Land manager: USDA Forest Service, Lewis and Clark National Forest.
Material: Fossils, calcite crystals.
Tools: Rock hammer.
Vehicle: Any.
Accommodations: Camping and RV parking in Lewis and Clark National Forest; camping, RV parking, and motels in Great Falls.
Special attractions: None.
Finding the site: From Augusta at the junction of Main Street and Manix Street, drive 3.5 miles west on Manix Street and turn right onto Sun River Road toward Sun River Canyon. Drive 15.2 miles and park just after the canal. The road cut is on the left and the road has to be crossed. *DeLorme: Montana Atlas & Gazetteer:* Page 69 D7.

Rockhounding

Through a series of faults, the overthrust belt of the Sawtooth Mountains has placed Mississippian Madison limestone above younger Cretaceous shales and sandstones. The Sun River flows through a canyon, exposing the various ages of the strata. Nice specimens of horn coral and brachiopods can be found in the Madison limestone throughout the canyon area. The limestone is extremely difficult to break and the best fossil collecting may be found by exploring the loose pieces of rock along the base of the road cut that has already weathered from the cliffs. Think of the particular patch mentioned as a starting point.

Belt

See map on page 40
Land type: High hills, road cut.
GPS: N. 47.37944 / W. 110.92342
Best season: Spring through fall.
Land manager: Montana DOT.
Material: Fossils.
Tools: Rock hammer, chisel.
Vehicle: Any.
Accommodations: Camping and RV parking in Lewis and Clark National Forest; camping, RV parking, and motels in Great Falls.
Special attractions: The C.M. Russell Museum in Great Falls displays art by the cowboy artist along with exhibits on the American West. The Lewis and Clark National Historical Trail Interpretive Center in Giant Springs Heritage State Park details the history of the Lewis and Clark expedition with special emphasis on the Montana leg of the journey. The Giant Springs Fish Wildlife and Parks Visitor Center and Fish Hatchery just outside of Great Falls is located on one of the largest freshwater springs in the world.
Finding the site: From the junction of U.S. Highway 87/Montana Highway 200 and U.S. Highway 89 east of Great Falls, drive north on the road to Belt for 2.7 miles to a large stretch of road cuts on the right. Watch for speeding traffic and park well off the road. *DeLorme: Montana Atlas & Gazetteer:* Page 57 A7.

Rockhounding

Jurassic coal seams around the area of Belt tell that once a more tropical time existed in the area that is now Montana. The seams resulted from thick deposits of organic matter in marshy low-oxygen environments that became buried by sand and mud deposited above and below them. Over time the material became coal seams. In 1893 Belt became home to the first commercial coal mine in Montana, and mining of the many coal deposits continued until the 1950s.

Today, evidence of the changing environment can be collected around the area of Belt. Leaf and other plant impressions can be found in the layers of shale. The best samples are collected by splitting the dark gray shale of the road cut using a hammer and chisel. It could take some time to find a complete sample but, due to the detailed quality of the impressions, it may be worth the energy.

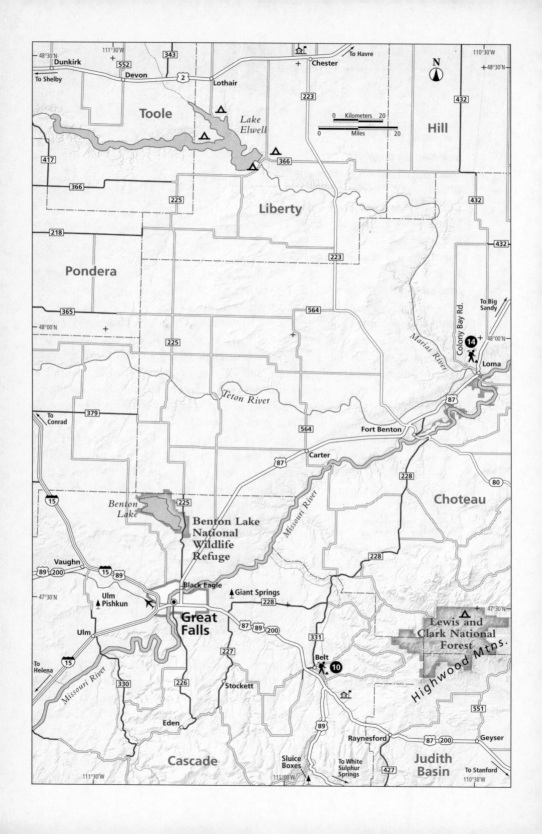

Superior

See map on page 42
Land type: Mountains.
GPS: N. 47.24050 / W. 114.85509
Best season: Spring through fall.
Land manager: USDA Forest Service, Lolo National Forest.
Material: Pyrite, arsenopyrite, quartz.
Tools: None.
Vehicle: Any.
Accommodations: Camping and RV parking nearby in Lolo National Forest; camping, RV parking, and motels in Superior.
Special attractions: The Mineral County Museum in Superior has displays on mining and local history.
Finding the site: From Interstate 90 in Superior, take exit 47 north about 0.5 mile and turn left onto Mullan Road. From Mullan Road, turn north onto Flat Creek Road. Drive 2 miles to a fork in the road, at which point you stay left and continue 1.5 miles to large piles of mine tailings on the left side of the road. *DeLorme: Montana Atlas & Gazetteer:* Page 52 A3.

Rockhounding

Considerable quantities of zinc, lead, copper, silver, and some gold have been recovered from the Superior/St. Regis area of Montana during the past 100 years. The ores were found in veins associated with several major fault zones. The majority of the rocks occurring here are metamorphic types of the Precambrian Belt series. Some younger igneous intrusions also occur throughout the area. Many of the mineral deposits that have been mined can be traced into the Coeur d'Alene District in Idaho just to the west.

The samples of pyrite and arsenopyrite at Superior are mediocre for collections due to weathering. I didn't spend much time at this site, and there may be more minerals waiting to be discovered. Most of the tailings piles do have nice pieces of quartz large enough to cut or sphere. Tailings occur throughout the area, so don't feel limited to exploring just the suggested locality. Farther up the road from the site mentioned there are several more piles, along with a warning about a mysteriously bright orange arsenic-bearing "toxic creek."

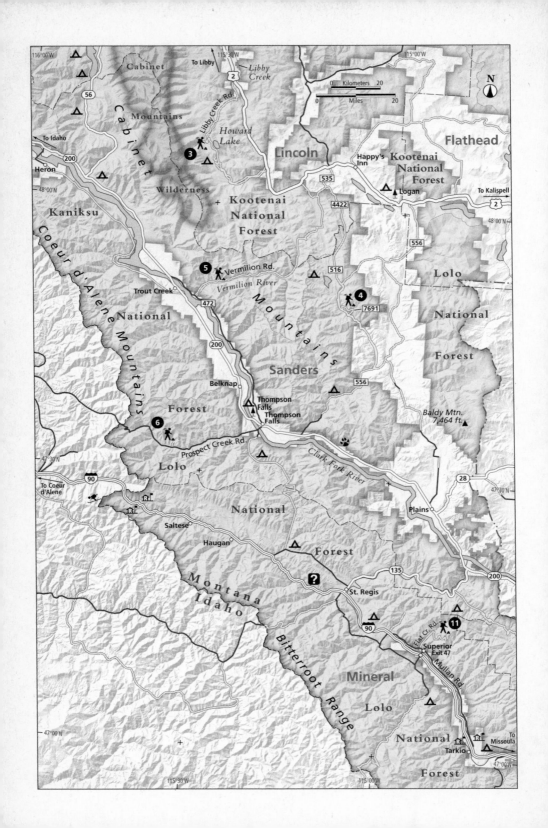

Searching tailings piles of Superior.

Of course, it is recommended that you stay out of the creek. Like many old mining sites, the charm and history come along with a significant cost to the environment.

Kalispell

See map on page 45
Land type: Mountains.
GPS: N. 48.56519 / W. 114.67420
Best season: Late spring through fall.
Land manager: USDA Forest Service, Flathead National Forest.
Material: Limonite after pyrite cubes, pyrite.
Tools: Rock hammer.
Vehicle: Any.
Accommodations: Camping and RV parking in Flathead National Forest; camping, RV parking, and motels in Kalispell.
Special attractions: Flathead Lake and Glacier National Park. The Northwest Montana Historical Society Museum in Kalispell has displays on local history.

Various sizes of limonite after pyrite cubes.

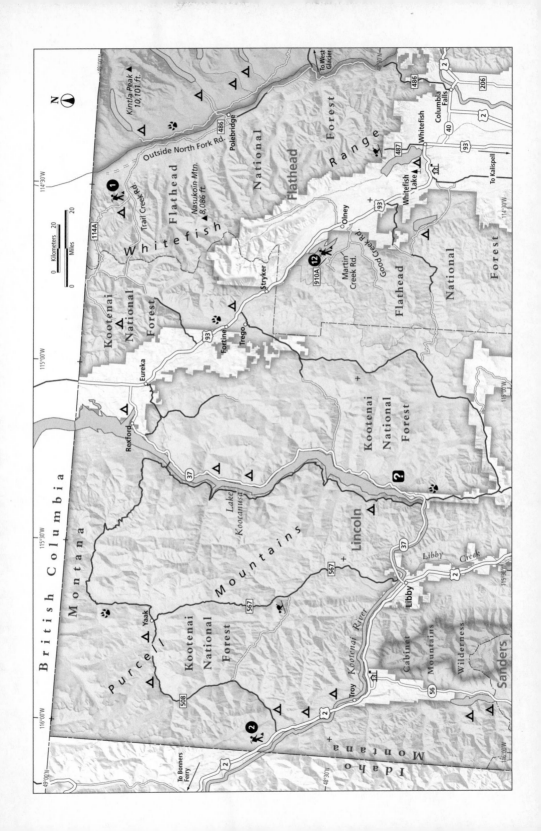

Finding the site: From U.S. Highway 93 north of Kalispell just before the town of Olney and about 0.5 mile after the Girl Scouts Camp, turn left onto Good Creek Road. Drive 3.3 miles and turn right onto Martin Creek Road. Drive 4.3 miles and park on the right at Forest Road 910A. If the gate is closed, walk 0.5 mile up FR 910A to the sharp corner and several outcrops of blue-green rock in the mountainside. Look for evidence of previous digging. *DeLorme: Montana Atlas & Gazetteer:* Page 82 C4.

Rockhounding

The Belt rock in this particular area produces some quality pseudomorphic limonite after pyrite cubes. Originally formed as pyrite, the limonite occurs as a secondary material due to oxidation of the original pyrite. Small samples of pyrite cubes can be found in the greenish blue rocks just before the bend; but the limonite after pyrite cubes, which are up to 1½ inches in size, are found in the rock outcrops just at, and slightly past, the corner. The outcrop has been highly worked, and it is easy to tell where to begin the search. The biggest cubes are found by breaking pieces loose from the parent rock and hoping for a lucky split. Small pieces can be found in the loose rocks at the bottom of the outcrop. The cubes tend to easily "pop" free from their parent rocks.

Craig

See map on page 36
Land type: Low mountains, road cut.
GPS: N. 47.15588 / W. 111.82723
Best season: Any.
Land manager: Cascade County.
Material: Stilbite, laumontite, mesolite, calcite, amethyst, and augite.
Tools: Rock hammer.
Vehicle: Any.
Accommodations: Camping and RV parking nearby on Holter Lake; camping, RV parking, and motels in Great Falls and Helena.
Special attractions: None.
Finding the site: From Interstate 15 north of Craig, take exit 244 (Canyon Access). Drive north on the frontage road (Recreation Road) for 0.5 mile to

Augite crystals found in volcanic rubble.

a large road cut on the right between the frontage road and I–15. Park in the turn out on the right side of the road and use caution while crossing the road. *DeLorme: Montana Atlas & Gazetteer:* Page 56 B2.

Rockhounding

Road cuts between Craig and Hardy on the frontage roads to I–15 yield specimens of the zeolite minerals stilbite, laumontite, and mesolite, as well as calcite, rare amethyst, and augite crystals. Zeolite is a term used to describe any number of minerals of the zeolite group, a group of hydrous aluminosilicates of the alkali or alkaline earth metals. There are several road cuts in the area that contain different minerals. At this particular site, very nice augite crystals can be found by searching the purple volcanic rubble leading up toward the interstate. Wherever you choose to rockhound, keep an eye out for rare specimens of amethyst in small vugs of the host rock.

Loma

See map on page 40
Land type: Hills.
GPS: N. 47.95092 / W. 110.53590
Best season: Spring through fall.
Land manager: Bureau of Land Management (BLM).
Material: Fossils, barite crystals.
Tools: Rock hammer.
Vehicle: Any.
Accommodations: Camping, RV parking, and motels in Great Falls.
Special attractions: Museum of the Northern Great Plains and Museum of the Upper Missouri River in Fort Benton have collaborative exhibits on the history of the area. The Upper Missouri National Wild and Scenic River Visitor Center has information on the wildlife and natural history of the river.
Finding the site: This site is rather small and complicated to locate; a BLM map is recommended. From the small town of Loma, just north of Fort Benton on U.S. Highway 87, drive to the north end of Loma and turn left onto Colony Bay Road. Drive about 1.5 miles down the road and turn right onto a small dirt BLM road. The first hill is on the left just 0.1 mile down the road, and the second is just 0.2 mile down the road. If the gate is closed, park off Colony Bay Road and walk into the BLM land by following the faint road. *DeLorme: Montana Atlas & Gazetteer:* Page 71 B4.

Rockhounding

Marine shales and sandstones of the Late Cretaceous Colorado Group appear predominately to the north and west of Great Falls, and prized fossil specimens from these rocks can be found, especially along the drainage basins of the Sun, Teton, and Marias Rivers.

This site can produce some wonderful specimens of marine fossils, especially ammonites and oysters—all of which retain their original mother-of-pearl shell. The fossils are found on either side of the BLM road. The best specimens are found by easily splitting the shale and loose shale concretions. Along with the marine fossils, nice barite crystals can also be found in the shale. The crystals tend to be about ¼-inch long and form in clusters.

Neihart

See map on page 51
Land type: Mountains.
GPS: N. 46.97486 / W. 110.69608
Best season: Summer through fall.
Land manager: USDA Forest Service, Lewis and Clark National Forest.
Material: Pyrite, malachite, sulphur.
Tools: Rock hammer.
Vehicle: High clearance.
Accommodations: Camping and RV parking in Lewis and Clark National Forest; camping, RV parking, and motels in White Sulphur Springs.
Special attractions: None.
Finding the site: From the junction of Montana Highway 200 and U.S. Highway 89 east of Great Falls, drive south on US 89 for 32.4 miles and turn left onto Forest Road 3323—just before you reach the town of Neihart. Drive about 3 miles down FR 3323, turn right at the fork in the road, and continue 0.2 mile to old mine tailings on the left side of the road.

Or, from White Sulphur Springs drive north on US 89 for 43 miles and turn right onto FR 3323, just past the town of Neihart, and follow the previous directions. *DeLorme: Montana Atlas & Gazetteer:* Page 57 C8.

Rockhounding

Discovered in 1881, the deposits near Neihart have produced close to $17 million worth of minerals. Though silver was the chief metal being mined here, lead, copper, gold, and zinc were also found. The veins and other ore bodies containing the minerals of these metals occur primarily in Precambrian metamorphic and sedimentary rocks along with younger igneous intrusions.

Mine tailings always seem to yield something of interest, but nothing too spectacular awaited me at this site at the extensive tailings piles of the Silver Dyke Mine. Nice specimens of malachite are easily found, along with sulphur, but they're of no exceptional quality. The mine tailings and the quality of the specimens do increase as you drive farther up the mountain, past the site, on a more difficult high-clearance jeep trail toward a large mine and abandoned buildings. The minerals and some interesting crystals are much more abundant here.

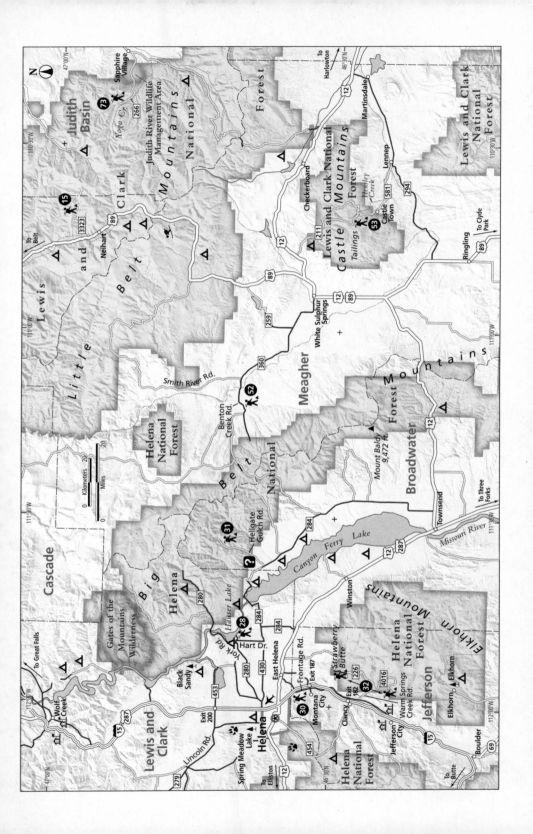

Dearborn River

See map on page 53
Land type: High hills, road cut.
GPS: N. 47.15955 / W. 112.21612
Best season: Spring through fall.
Land manager: Montana DOT.
Material: Stilbite crystals.
Tools: None.
Vehicle: Any.
Accommodations: Camping and RV parking in Helena National Forest; camping, RV parking, and motels in Lincoln.
Special attractions: None.
Finding the site: From the junction of Montana Highways 200/434 northeast of Lincoln, drive southeast on MT 434 for 3.8 miles to several hillside road cuts

Lincoln stilbite crystals.

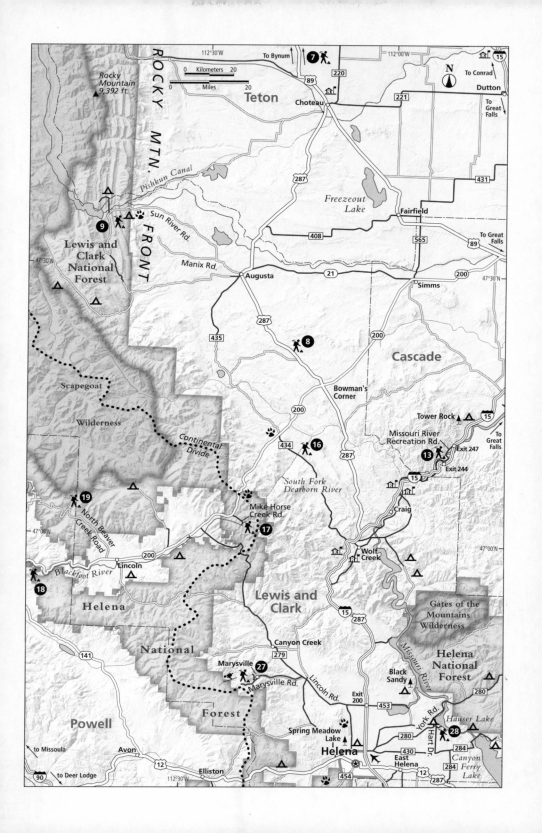

just past the South Fork Dearborn River. The river is seasonal and not labeled, so watch for the bridge. Just past the river and bridge, there is a hill on the left, and the crystals can be found weathering out on the hillside. *DeLorme: Montana Atlas & Gazetteer:* Page 55 B8.

Rockhounding

The sheaf-like stilbite, a zeolite mineral, weathers from the parent volcanic rock, and crystals of it can be found quite easily sifting through the crumbling hillsides near the river. The crystals tend to be stained a pink to orange color. They're usually about ¼ inch in size, but there are reports of the crystals being much larger, up to 1 inch.

Mike Horse Creek

See map on page 53

Land type: Mountains, stream.

GPS: N. 47.02756 / W. 112.35814

Best season: Spring through fall.

Land manager: USDA Forest Service, Helena National Forest.

Material: Pyrite, chalcopyrite, galena, sphalerite, malachite, azurite, quartz, jasper.

Tools: Rock hammer.

Vehicle: Any.

Accommodations: Camping and RV parking in Helena National Forest; camping, RV parking, and motels in Lincoln.

Special attractions: None.

Finding the site: From the junction of Montana Highways 200/434 northeast of Lincoln, drive 12.7 miles southwest on MT 200 and turn left onto Mike Horse Creek Road. Drive 2.2 miles, park, and explore the tailings piles along the creek. *DeLorme: Montana Atlas & Gazetteer:* Page 55 B8.

Rockhounding

This site was originally intended to lead to the tailings piles at the Mike Horse Creek Mine 3 miles farther up the road. At the time of my visit, however, there was a closed gate, and land access was questionable. Gold, silver, lead, and copper were mined from the Mike Horse, which was once the biggest mine of the Heddlestone District. The minerals are found as a result of igneous intrusions into the Precambrian overthrust belt.

Interestingly enough, the tailings piles 3 miles down the road from the mine, at this site on national forest land, produce adequate copper minerals available for collection. Large dumps litter the area, and specimens of pyrite, chalcopyrite, and malachite are easily found in the tailings piles along the creek toward the dam. Other minerals, such as galena, sphalerite, azurite, small quartz crystals, and jasper, require a bit more searching.

Blackfoot River

See map on page 53
Land type: River, gravel bars.
GPS: N. 46.93844 / W. 112.87714
Best season: Spring through fall.
Land manager: Bureau of Land Management (BLM).
Material: Agate, jasper, chalcedony.
Tools: None.
Vehicle: Any.
Accommodations: Camping and RV parking in Helena National Forest; camping, RV parking, and motels in Lincoln.
Special attractions: None.

Lincoln agate tends to be less translucent than the agate found in eastern Montana.

Finding the site: From Lincoln drive 12.3 miles west on Montana Highway 200 to BLM signs for the Blackfoot River fishing access on the left; the parking area is on the right. Enter the public fishing area and follow the trails to gravel bars and clearings in the plateaus along the river. BLM signs are very helpful in identifying the area open for public use. *DeLorme: Montana Atlas & Gazetteer:* Page 55 C5.

Rockhounding

At the time of my visit, access to this fishing area was behind a barbed-wire fence, and in typical rugged Montana fashion, BLM had built a steep double-sided ladder over the fence to allow public access instead of installing a gate. Once over the ladder, the agate is found in the gravel clearings, but it is not abundant. The agate is usually red, brown, gray, and green and not translucent like pieces from the eastern part of the state; most of it tends to look more like jasper with nice patterning.

Coopers Lake

See map on page 53

Land type: Mountains, road cut.

GPS: N. 47.04934 / W. 112.85589

Best season: Late spring through fall.

Land manager: USDA Forest Service, Helena National Forest.

Material: Pyrite.

Tools: Rock hammer, sledge, chisel.

Vehicle: Any.

Accommodations: Camping and RV parking in Helena National Forest; camping, RV parking, and motels in Lincoln.

Special attractions: None.

Finding the site: From Lincoln drive west on Montana Highway 200 for 1.5 miles and turn right onto North Beaver Creek Road (Forest Road 4106). Drive 15 miles to a large road cut of blue-green rock with a vehicle turnout. *DeLorme: Montana Atlas & Gazetteer:* Page 55 B5.

Rockhounding

Excellent pyrite cubes up to 2 inches in size can be found in the exposed blue-green Belt rock in this area. The cubes aren't overly abundant, and this location appears to be a popular spot with rockhounds. The best way to collect specimens is to break rock loose from the outcrop and hope to expose a pyrite cube. There are some loose pieces that can be found by searching downhill from the road cut, but be careful—it's very steep. Since the cubes are fairly rare, look for Belt rock with impressions from cubes that have been broken loose.

Rattler Gulch

See map on page 60
Land type: High hills, road cut.
GPS: N. 46.69647 / W. 113.23018
Best season: Any.
Land manager: Montana DOT.
Material: Fossils.
Tools: Rock hammer.
Vehicle: Any.
Accommodations: Camping and RV parking in Lolo National Forest; camping, RV parking, and motels in Deer Lodge and Missoula.
Special attractions: Garnet ghost town is located up Bear Gulch about 10 miles north of the Interstate 90 frontage road. There are accessible tailings piles just before Garnet on the left side of the road—but aside from pyrite and quartz crystals there is little to be found.
Finding the site: From Drummond on I–90 take exit 153 to the north side of the freeway. Drive 4 miles northwest on the frontage road and turn left at the sign for Rattler Gulch. Drive a short distance, about 0.1 mile, through a very "colorful" tunnel under the freeway decorated personally by local teenage artists. Turn right onto the gravel road that follows the railroad tracks and drive 0.2 mile to vertical layers protruding from a hill on the right. *DeLorme: Montana Atlas & Gazetteer:* Page 54 D3.

Rockhounding

These steeply tilted rock layers are sedimentary deposits of the Jurassic and Cretaceous age. The marine layers of the Jurassic rocks have produced some fairly well-preserved fossils. Several species of mollusks have been reported from the shales and limestones present near the east end of the railroad cut. The oyster *Gryphaea* is one of the more common fossils to be found here. The rock breaks easily, and weathered specimens can be found already broken free from the outcrop.

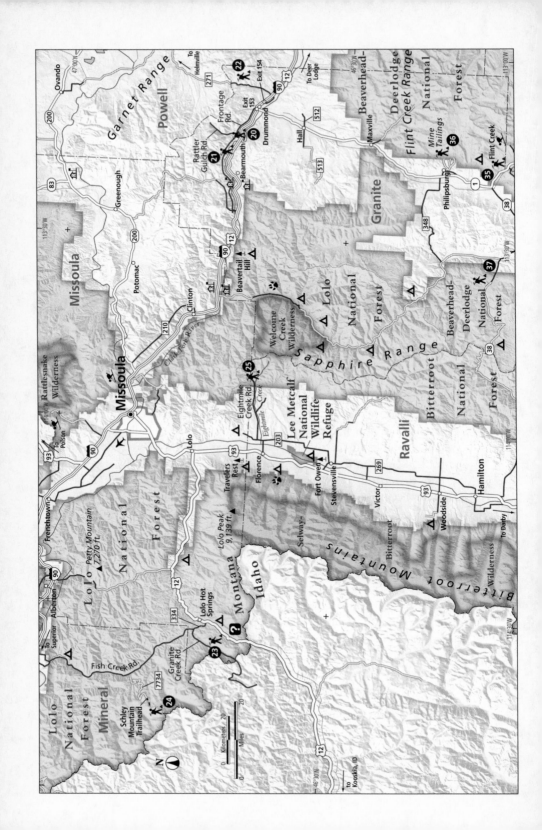

Clark Fork River

See map on page 60
Land type: Mountains, road cut.
GPS: N. 46.71689 / W. 113.32320
Best season: Spring through fall.
Land manager: Granite County.
Material: Calcite crystals.
Tools: Rock hammer.
Vehicle: Any.
Accommodations: Camping and RV parking in Lolo National Forest; camping, RV parking, and motels in Deer Lodge and Missoula.
Special attractions: Garnet ghost town is located up Bear Gulch about 10 miles north of the Interstate 90 frontage road. There are accessible tailings piles just before Garnet on the left side of the road—but aside from pyrite and quartz crystals there is little to be found.
Finding the site: From Drummond on I–90 take exit 153 to the north side of the freeway. Drive northwest on the frontage road for about 8 miles to the slopes between the old highway (frontage road) and the Clark Fork River. Nice samples of calcite crystals can be collected from the top of the hill down to the river. There are some excellent specimens of calcite crystals in the area but caution must be exercised—the slope is very steep and the rocks on it are loose. It is recommended that you stick to collecting samples on the top. *DeLorme: Montana Atlas & Gazetteer:* Page 54 D3.

Rockhounding

Very nice samples of calcite crystals can be found in the reddish and pink country rock on the hillside. Samples of small clusters of crystals are already exposed and can be easily collected. Most specimens of the attractive clusters can't be removed from the rock without damaging it, but they are still quite nice. The clusters are large, and the crystals themselves are usually small, about half an inch in length.

A cluster of small calcite crystals.

Helmville

See map on page 60
Land type: Hills, road cut.
GPS: N. 46.71689 / W. 113.32320
Best season: Summer through fall.
Land manager: Montana DOT.
Material: Fossils.
Tools: Rock hammer, trowel.
Vehicle: Any.
Accommodations: Camping and RV parking in Lolo National Forest; camping, RV parking, and motels in Deer Lodge and Missoula.
Special attractions: Garnet ghost town is located up Bear Gulch about 10 miles north of the Interstate 90 frontage road. There are accessible tailings piles just before Garnet on the left side of the road—but aside from pyrite and quartz crystals there is little to be found.
Finding the site: From I–90 in Drummond, take exit 154 to the south side of the highway. Drive southeast on the frontage road for 1.5 miles and turn left onto the road to Helmville (Montana Highway 271). Travel 6.5 miles to a large road cut on the right side of the road, a little more than 0.5 mile from where the pavement ends. The site is difficult to find since the rock containing the fossils is hard to see. Keep an eye out for the thin light-colored shale eroding from several hills facing the road. *DeLorme: Montana Atlas & Gazetteer:* Page 54 D3.

Rockhounding

Extremely well-preserved fossilized leaves can be found in mid-Tertiary rocks northeast of Drummond. Insects and a few rare fish skeletons have been reported in the area as well. The light-colored shale on the right side of the road contains the fossils. Nice specimens are not overly abundant, so you will have to look hard to find anything. Diligence can pay off, however, because the leaves tend to be finely detailed. Be sure to bring sufficient packaging materials to wrap anything collected. The brittle shale breaks easily.

Lolo Hot Springs Area

See map on page 60

Land type: Mountains, road cut.

GPS: N. 46.69818 / W. 114.60468

Best season: Late spring through fall.

Land manager: USDA Forest Service, Lolo National Forest.

Material: Quartz crystals.

Tools: Rock hammer, shovel, ¼ inch screen.

Vehicle: Any.

Accommodations: Camping and RV parking in Lolo National Forest; camping, RV parking, and motels in Lolo and the resort town of Lolo Hot Springs.

Special attractions: Lolo Hot Springs is a natural source of geothermal water. It has been developed into a resort and provides weary vacationers with relaxation and comfortably warm water in a well-kept pool.

Finding the site: From the junction of U.S. Highways 93/12 in Lolo, drive west on US 12 for about 25 miles. Go 1 mile past the Lolo Hot Springs resort area and turn right onto Forest Road 343. Drive 2 miles along FR 343 and turn left onto Granite Creek Road. Drive 5 miles down to the site; look for evidence of digging in the road cuts. *DeLorme: Montana Atlas & Gazetteer:* Page 52 D4.

Rockhounding

In the area north and west of Lolo Hot Springs, along Granite Creek and its tributaries, numerous outcrops of the granitic Idaho batholith are accessible because of extensive logging. Beautiful crystals of clear to very dark smoky quartz can be found in miarolitic cavities within the rock and even loose in the soil. Their occurrence and digging for them in the decomposed granitic soil is somewhat similar to Crystal Park.

The directions mentioned will take you to a general location where crystals can be found. Most of the crystals in this area are found loose in the soil using a screening method. It is generally productive to work on a hole where others have previously been digging. To try your luck at scouting out a new location, look for soil similar to the digging area near the road and use a screen. In order to locate other areas where crystals may occur, especially in hard rock, carefully examine the rock exposures in road cuts, hillsides, and exposed areas of recent logging. In particular, look for patches of smoky quartz or the presence of unusually large grains of quartz or feldspar (pegmatite). Most of the rock in the area is generally uniform in texture, with relatively small grains of quartz and feldspar, and so the pegmatite nature of a rock outcrop may indicate a potential source of well-formed crystals. The area of granite to be explored spans for miles and is in no apparent structure. The crystals may occur throughout the forest.

Extreme care should be taken in extracting the crystals. Many beautiful specimens have been destroyed by hasty decisions and poor judgment in removing them from the host rock. Groups of brilliant crystals of various sizes, ranging from pounds heavy to less than an inch big, can be the prize of a patient collector. The crystals uncovered from the loose soil may be smaller, usually less than an inch in size, but the work is also easier and gem quality can still be found.

Snowbird Mine

See map on page 60
Land type: Mountains.
GPS: N. 46.77663 / W. 114.79389
Best season: Summer.
Land manager: USDA Forest Service, Lolo National Forest.
Material: Quartz crystals, fluorite, parisite (a rare earth carbonate).
Tools: Rock hammer.
Vehicle: Any.
Accommodations: Camping and RV parking in Lolo National Forest; camping, RV parking, and lodge in Lolo and resort town of Lolo Hot Springs.
Special attractions: Lolo Hot Springs is a natural source of geothermal water. It has been developed into a resort and provides weary vacationers with relaxation and comfortably warm water in a well-kept pool.
Finding the site: From the junction of U.S. Highways 93/12 in Lolo, drive west on US 12 for 25.5 miles. Just past the Lolo Hot Springs resort area, turn right onto Fish Creek Road (Forest Road 343). Take Fish Creek Road for 13.8 miles and turn left onto the road leading toward the Schley Mountain Trailhead. Drive 11.7 miles, stay left at the fork, and drive toward the upper trailhead, located 2.9 miles past the fork. At the time of my visit, the road which leads to the mine from this point was gated off and closed to motorized traffic past the trailhead. You may have to continue up this road by foot; park at the trailhead and walk about 1 mile farther on the road, not on the hiking trail. It is an uphill hike to the mine once you pass the gate. When the old mining road splits about halfway to the mine, stay on the low (right) path. Many of the forest roads leading to Snowbird Mine do not open until mid- to late summer, so always inquire locally about accessibility. *DeLorme: Montana Atlas & Gazetteer:* Page 52 C4.

Rockhounding

About seventy-seven million years ago, a carbonatite sill intruded along a fault in sedimentary Belt series rocks. The Snowbird Mine developed the fluorite associated with the fault's sill. The large deposits appear to have been mined for quite some time, but today the mine is long since defunct.

For the rockhound this abandoned mine is a treasure trove. The old road

Large quartz crystals at the blasted areas of Snowbird Mine.

to the mine sparkles with scattered fluorite like a trail of sprinkled diamonds. Once at the mine, high up in the mountain, the steep hike is rewarded with a spectacular view of some of Montana's most lush territory. Veins of fluorite were mined by blasting the mountainside, and several large caves exist as scars to the story. The fluorite veins inside the caves are bright shades of green, white, and a deep blackish purple. Even though acres of dumps are located beneath the mine, some of the best samples can be collected from the source in the caves and on the mountainside.

Large chunks of quartz blasted from the mountain rest in piles along the edges of the caves. Translucent quality quartz crystals occur within some of these rocks up to several inches long, but it's unlikely they can be removed without damaging the crystal. Low-grade quartz crystals without any translucency occur up to a weight of a couple hundred pounds. These behemoths are often up to a foot in diameter, and they're impossible to remove without damage

Fluorite caves at Snowbird Mine.

to the crystal. The beautiful fairy-tale size of the crystals is best viewed and left untouched for other rockhounds to enjoy. Any attempt at removing the crystals would only result in damaging their beauty and most likely your tools. Take pictures and preserve the beauty for another to enjoy.

Parisite crystals up to 9 inches long have been found at the Snowbird Mine. At the time of visit to the mine, diligent searching of the loose rock for the brownish crystals produced only small samples.

Eightmile Creek

See map on page 70
Land type: Mountains, creek.
GPS: N. 46.65137 / W. 113.88883
Best season: Late summer through fall.
Land manager: USDA Forest Service, Bitterroot National Forest.
Material: Gold.
Tools: Gold pan, shovel, fleck flask, bucket.
Vehicle: Any.
Accommodations: Camping and RV parking in Bitterroot National Forest; camping, RV parking, and motels in Lolo and Missoula.
Special attractions: None.
Finding the site: From the small town of Florence just south of Lolo, drive east from U.S. Highway 93 on Eastside Highway for 1.5 miles and turn left onto Montana Highway 203. Drive 0.7 mile and turn right onto Eightmile Creek Road. Continue 3.2 miles to a fork and stay left. The national forest area is 3.7 miles farther down Eightmile Creek Road. The creek alongside the road produces gold from the entrance to the forest area up to a little past a bridge. *DeLorme: Montana Atlas & Gazetteer:* Page 53 D8.

Rockhounding

In the area around Eightmile Creek, sedimentary Belt series rocks are intruded upon by the Idaho batholith. Along the layer of contact between the intrusive granite and the porphyry, the now-exposed minerals were deposited. Mining of these minerals in the Eightmile District began in the late 1800s and must not have been entirely profitable as very little information on ore production is available. The mining was on a small scale and was mostly placer mining. The area along Eightmile Creek in the national forest has been dredged, but don't be discouraged, the hillsides produce enough color to keep a recreational panner happy. You'll just have to dig from the nearby hillsides and road cuts and carry the dirt to the creek.

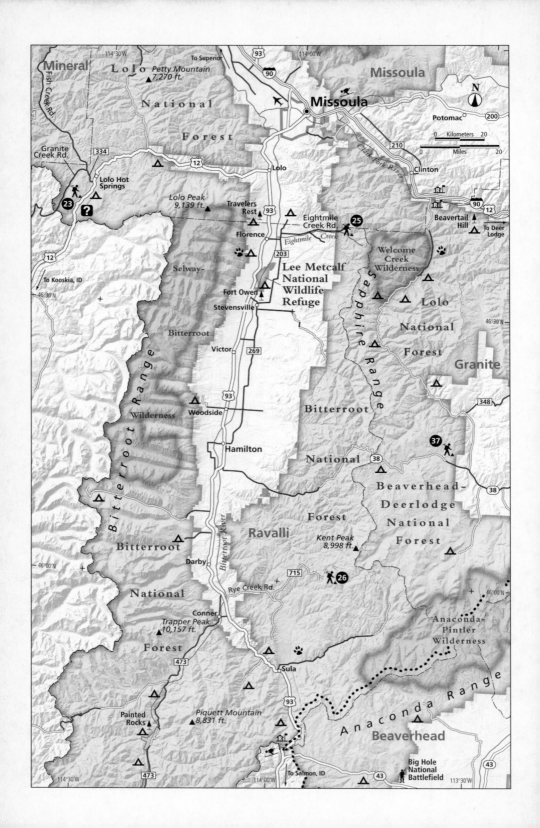

Darby Mine

See map on page 70
Land type: Mountains.
GPS: N. 46.01088 / W. 113.88063
Best season: Summer through fall.
Land manager: USDA Forest Service, Bitterroot National Forest.
Material: Fluorite.
Tools: None.
Vehicle: High clearance.
Accommodations: Camping and RV parking in Bitterroot National Forest; camping, RV parking, and motels in Hamilton.
Special attractions: None.
Finding the site: From Darby drive south on U.S. Highway 93 for 4 miles and turn left onto Rye Creek Road. Drive 7.7 miles and turn left onto Forest

Several buildings of the old Darby Fluorite Mine still remain.

Road 715. Stay on FR 715 for 13 miles to a fork, turn right, and go 0.2 mile to the old Darby mine. There is a gate on this road that may be closed. If so, park and walk the last 0.1 mile to the tailings. *DeLorme: Montana Atlas & Gazetteer:* Page 37 D8.

Rockhounding

Several deposits containing the mineral fluorite are located in the area east of Darby. The fluorite is high grade and occurs as white, green, and deep purple masses. Sometimes all of the colors occur as a very colorful chunk for collectors. The grains are relatively small, less than 1 inch in diameter, making it quite unlike the popular cleavable fluorite from Illinois.

The fluorite deposits occur in igneous and metamorphic rocks and may be related to the Idaho batholith in origin. The mineral is found as lenses within the surrounding rock and was exposed in relatively large amounts at the surface. Mining did not begin until the 1950s and continued until recently. As always, watch out for private property and trespassing signs and stay away from anything that may be private.

Marysville Mine Tailings

See map on page 74
Land type: Mountains.
GPS: N. 46.73434 / W. 112.32517
Best season: Late spring through fall.
Land manager: USDA Forest Service, Helena National Forest.
Material: Pyrite, quartz crystals, sphalerite, galena, azurite.
Tools: Rock hammer.
Vehicle: High clearance.
Accommodations: Camping and RV parking in Helena National Forest; camping, RV parking, and motels in Helena.
Special attractions: Helena is Montana's state capital, and the capitol alone is worth a visit. Construction began in 1899, and the building is faced with native sandstone and granite, with a dome made of Montana copper. Across the street from the capitol is the Montana Historical Society's museum, library, and archives. The museum displays some of the early mining equipment and ores found at nearby mining camps.

In 1864 placer gold made Last Chance Gulch, now a main street of Helena, one of the most famous places in Montana. The gold in the gravels of the gulch originated in the lode deposits that occurred along Last Chance Creek and its tributaries. It was sought using a variety of methods from simple panning to large-scale dredging. By 1910 the gravels of the gulch had yielded about $16 million in gold.

Finding the site: From Interstate 15 just north of Helena, take exit 200, Lincoln Road/Montana Highway 279. Drive northwest on Lincoln Road for 9.5 miles and turn left onto Marysville Road. Drive 6.8 miles, passing through the town of Marysville, and turn left onto Ottawa Gulch Road. Drive 1.3 miles to a few tailings piles and old mines on both sides of the road. *DeLorme: Montana Atlas & Gazetteer:* Page 55 D8.

Rockhounding

Placer gold deposits were first discovered around Marysville in the 1860s, but true profit came in the 1800s when hard-rock mining of the edges of granitic intrusions associated with the Boulder batholith produced the once major gold-mining town of Marysville. Marysville and the nearby towns of Rimini, Austin, Wilborn, and Lincoln yielded quantities of gold valued in the millions.

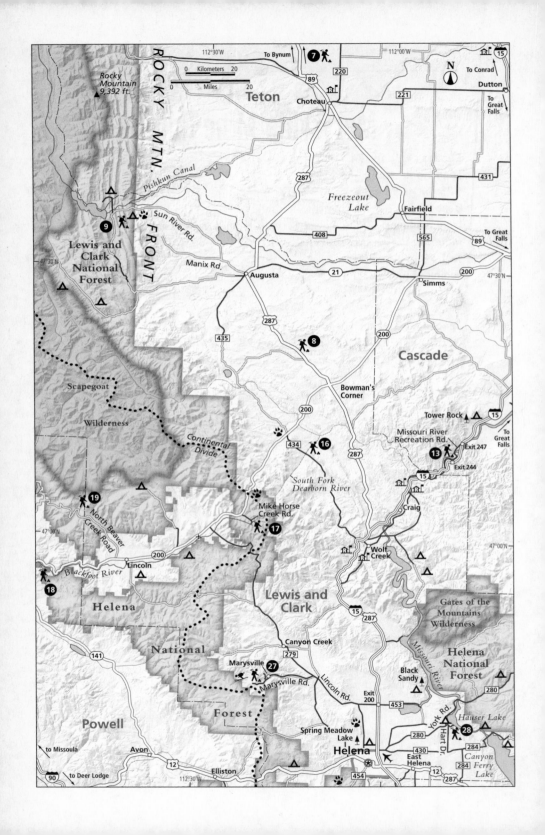

These are interesting places to visit, and the stream gravels often show gold for the prospector. Area mine dumps and tailings have undergone considerable weathering, and mineral specimens of fair size may be very difficult to find. Diligent investigating may uncover specimens of micromount or thumbnail-collection quality.

Pyrite, quartz crystals, sphalerite, galena, azurite, and other minerals associated with hard-rock mining are documented in the tailings piles of the Ottawa District area outside of Marysville. Exactly where they occur is debatable. The mine tailings at the time and location I visited provided only mediocre samples of quartz crystals and pyrite. Nevertheless, Marysville is an interesting place to visit and, with the assistance of a good forest map, the entire district could be explored.

Spokane Bar Sapphire Mine

See map on page 77
Land type: Hills.
GPS: N. 46.66776 / W. 111.83009
Best season: Open 9:00 A.M. to 5:00 P.M. seven days a week from late spring through early fall; winter hours vary.
Land manager: Private, Spokane Bar Sapphire Mine.
Material: Sapphire, garnet, ruby, topaz, moonstone, citrine, quartz, agate, jasper, and gold.
Tools: Most tools provided, or bring your own. Tweezers are recommended.
Vehicle: Any.
Accommodations: Camping and RV parking nearby on Canyon Ferry Reservoir; camping, RV parking, and motels in Helena.
Special attractions: The state capitol in Helena, which was constructed in 1899, is worth a visit. It is faced with native sandstone and granite, with a dome made of Montana copper. Across the street from the capitol is the Montana Historical Society's museum, library, and archives. The museum displays some of the early mining equipment and ores found at nearby mining camps.
Finding the site: From Helena take Montana Highway 280 (York Road) east toward the Missouri River/Canyon Ferry Lake for 8 miles. Turn right onto Hart Drive and after 0.2 mile turn left onto Castles Road. Drive 1 mile to the mine on the right. *DeLorme: Montana Atlas & Gazetteer:* Page 56 D2.

Rockhounding

Sapphires were first discovered in 1865 in terrace gravel deposits along the Missouri River northeast of Helena that were being worked for placer gold. They were a byproduct that clogged the riffles in sluice boxes and dredges, but when their value was recognized, mining the sapphires became almost as important as mining the gold. The development of synthetic corundum virtually eliminated the need for natural corundum as a source of abrasives, and commercial mining today is primarily restricted to gem-quality stones.

The Spokane Bar Sapphire Mine is situated on a high gravel terrace about 5 miles downstream from Canyon Ferry Dam and west of Hauser Lake. The mine is open year-round to the public for fee-digging for sapphires. All equipment

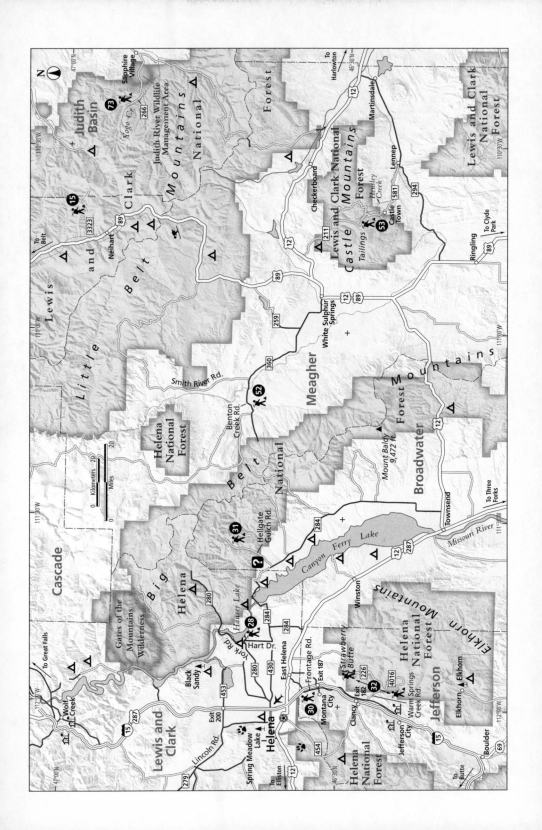

Chris Hodges digs his own sapphire gravel from the pits at the Spokane Bar Sapphire Mine.

needed is available for use at no additional cost. The gravels and digging area is confined to virgin or previously unprocessed ground, which will enhance one's chances of finding sapphires.

Buckets of prescreened gravel may be purchased if one chooses not to dig. If prescreened material is purchased, digging may be done for a small extra fee. The prescreened bags seem to give the best quantity and quality of sapphires and garnets, but seasoned rockhounds may enjoy digging for their treasures. For those interested in large quantities of sapphires, additional buckets for full-day "runs" come at a small cost. Customers screen their own dug gravels on-site and, once screened, the mine provides washtubs for rinsing and large tables for sorting the rinse material. There is also an office and rock shop on-site.

Whether you choose to dig or not, there are many treasures to be found screening the Spokane Bar's gravels. Almost every naturally occuring color of sapphires are found here. Most are light green-blue, about half a carat, and often of gem quality. The largest sapphire on record found at the mine was 155 carats; the largest gem-quality sapphire found was 50 carats. Anything you find in your purchased gravels at the mine is yours to keep; the rock shop proudly displays an article on the wall about a customer who found a sapphire at the mine worth several thousand dollars!

Other treasures are found as you wash and pick through the gravel. Tiny samples of ruby, garnet, topaz, moonstone, citrine, quartz, agate (often moss agate), and jasper are commonly found, and many pieces are of lapidary quality. Customers are also welcome to pan the excess dirt of their purchases for the fine gold that exists in the Spokane Bar terrace gravels.

For customers who find gem-quality sapphires, contact information for companies who offer faceting and heat-treating services through the mail is available through the rock shop. The bags of potential sapphire-bearing gravel that are for purchase are also available to be shipped, via mail, worldwide.

To write for a brochure or inquire about purchasing a bag of gravel, contact Spokane Bar Sapphire Mine, Russ and Deb Thompson, 5360 Castles Drive, Helena, MT 59602; (406) 227–8989, (877) DIGGEMS; www.sapphire mine.com.

Elliston

See map on page 81
Land type: Mountains.
GPS: N. 46.41962 / W. 112.51122
Best season: Summer through fall.
Land manager: USDA Forest Service, Helena National Forest.
Material: Agate nodules.
Tools: Rock hammer.
Vehicle: Any.
Accommodations: Camping and RV parking on-site in Helena National Forest; camping, RV parking, and motels in Helena.
Special attractions: None.
Finding the site: From U.S. Highway 12 in Elliston, drive east 0.5 mile, turn right onto Forest Road 227, and drive 13.5 miles to the Kading Campground. Park at the farthest end of the campground and follow the trail at the end of the road to the hilltops of the surrounding mountains. *DeLorme: Montana Atlas & Gazetteer:* Page 39 A7.

Rockhounding

Halfway between Helena and Deer Lodge is the last place you would expect to find nice specimens of agate—and only the trained eye can spot the cobble-like agate geodes in this area. Even then, a 1.5-mile strenuous uphill hike is required to reach the locality.

The agate nodules are quite small, usually a little larger than 1 inch in diameter, and difficult to see. Specimens are found loose in the soil and can occur along the trail, but the most concentrated locations are on

Agate nodules from Elliston.

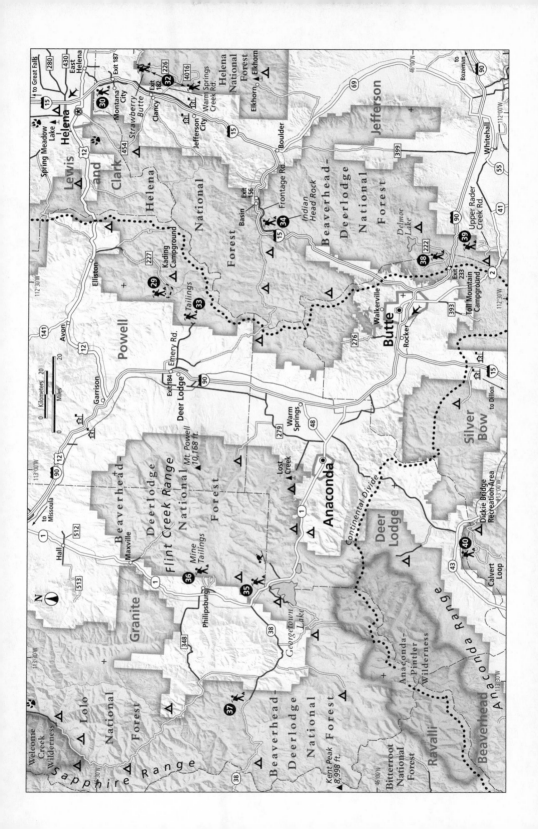

the very top of the mountains. Split nodules are commonly found and seem to bunch up in concentrations with solid specimens. The nodules contain agate in solid colors of white and gray and banded agate in white, gray, brown, and black. The specimens can be worth the hike though, as many of them contain geode-style quartz crystals inside banded agate similar to the Montana dryhead agate.

Montana City

See map on page 77
Land type: Mountains, road cut.
GPS: N. 46.54285 / W. 111.94830
Best season: Any.
Land manager: Jefferson County.
Material: Jasper, agate.
Tools: Rock hammer.
Vehicle: Any.
Accommodations: Camping, RV parking, and motels in Montana City and Helena.
Special attractions: The state capitol building in Helena, which was constructed in 1899, is worth a visit. It is faced with native sandstone and granite, with a dome made of Montana copper. Across the street from the capitol is the Montana Historical Society's museum, library, and archives. The museum displays some of the early mining equipment and ores found at nearby mining camps.
Finding the site: From Interstate 15 just south of Helena, take exit 187 (Montana City) and go to the frontage road on the west side of the interstate. The road cut is just north of the exit on the left side of the frontage road. It is best to park your car at the gas station that is located immediately across the frontage road when you exit the freeway and walk the short distance north to the first small, humble hill on the left. *DeLorme: Montana Atlas & Gazetteer:* Page 40 A2.

Rockhounding

The plume jasper to be found in this little hill is rather impressive. Beautiful large chunks, many several inches in diameter, are visibly eroding away from the cut. The stunning jasper occurs in several shades of brown, red, and a nice caramel color. With some luck and time, you can also find interesting samples of attractive plume agate.

Argo Mine

See map on page 77
Land type: Mountains.
GPS: N. 46.69100 / W. 111.58540
Best season: Late spring through fall.
Land manager: USDA Forest Service, Helena National Forest.
Material: Malachite, azurite, pyrite, quartz.
Tools: Rock hammer.
Vehicle: High clearance.
Accommodations: Camping and RV parking on Canyon Ferry Lake and Helena National Forest; camping, RV parking, and motels in Helena.
Special attractions: The state capitol building in Helena, which was constructed in 1899, is worth a visit. It is faced with native sandstone and granite, with a dome made of Montana copper. Across the street from the capitol is the Montana Historical Society's museum, library, and archives. The museum displays some of the early mining equipment and ores found at nearby mining camps.
Finding the site: From Helena drive east on Montana Highway 284/430 to Canyon Ferry State Park. Cross over to the east side of the lake and continue south toward the Hellgate Campground, 7.5 miles past the dam. Turn left onto Hellgate Gulch Road and drive 5.6 miles to large tailings piles on the right side of the road. A high-clearance vehicle is needed to cross the dirt mounds the Forest Service has placed continuously throughout the last 5 miles of Hellgate Road (for erosion control). If you decide to walk, however, it is a nice road with extremely beautiful scenery. *DeLorme: Montana Atlas & Gazetteer:* Page 56 D3.

Rockhounding

The sight of the great towering limestone cliffs of Hellgate Canyon is worth the visit alone. Monstrous ledges wall the narrow windy road, mountain goats traverse the high hills, and natural beauty is complimented by the Argo Mine 5 miles farther down the road, which has some of the best cutting-quality malachite to be found in the state.

The Argo copper mine of the Hellgate Mining district was operated in the early 1900s. By the end of the mine's short run, it had produced almost three million pounds of copper mined from fissures in the shale.

Samples of pyrite, quartz, azurite, and malachite can be found in the massive tailings piles surrounding the Argo Mine. There is plenty of material to collect just by walking along the base of the towering piles and picking up loose samples. This site is also home to some of the nicer samples of malachite and azurite of Montana.

Malachite from the Argo Mine.

Clancy and Strawberry Butte

See map on page 77
Land type: Mountains, trail.
GPS: N. 46.47652 / W. 111.91884
Best season: Late spring through fall.
Land manager: USDA Forest Service, Helena National Forest.
Material: Topaz, arsenopyrite.
Tools: Rock hammer, sledge hammer, chisel, and gloves.
Vehicle: Any.
Accommodations: Camping and RV parking in Helena National Forest; camping, RV parking, and motels in Helena.
Special attractions: The state capitol building in Helena, which was constructed in 1899, is worth a visit. It is faced with native sandstone and granite, with a dome made of Montana copper. Across the street from the capitol is the Montana Historical Society's museum, library, and archives. The museum displays some of the early mining equipment and ores found at nearby mining camps.
Finding the site: From Interstate 15 south of Helena, take Clancy exit 182 to the frontage road on the east side of the freeway. Drive south on the frontage road for 1 mile and turn left onto Warm Springs Creek Road. Drive 3.7 miles, turn right on Forest Road 4016, and continue 1.9 miles to the tailings piles and park.

To reach Strawberry Butte, follow the above directions to Warm Springs Road. Drive on Warm Springs Creek Road for 8.2 miles to Forest Road 226, the road on the left that leads to the Strawberry Fire Lookout. Take this road 0.2 mile and park at the gate. You will have to walk the rest of the way up toward the lookout. It is only about 0.5 mile to the first outcrop of topaz-containing rhyolite on the left side of the trail, and the outcrops continue up to the sharp bend to the right just before the lookout. *DeLorme: Montana Atlas & Gazetteer:* Page 40 A2.

Rockhounding

This is an interesting area because there are not only tailings piles to pick through but also a site to split for topaz.

The arsenopyrite is found in the tailings piles of the old mines. It tends to

An exposed pocket of Clancy arsenopyrite in quartz.

be rather interesting, and nice specimens of arsenopyrite crystal clusters can be found in cavities of the quartz.

The topaz is found in volcanic rock from a separate area. Topaz occurrences are generally related to tin–bearing veins of the igneous rock, like those which occur just south of Clancy above Boulder. The topaz crystals found north of Clancy on Strawberry Butte are usually less than 1 inch long and average only half an inch in length. The translucent prismatic crystals are most often clear to yellow in color, but reports of blue, lavender, and orange colors have been documented. The topaz formed in gas cavities within the light gray rhyolite, and splitting the rock to expose a cavity is a matter of time and luck. Keep in mind that topaz crystals have a tendency to fade after exposure. The crystals are not easy to find. To do so you must spend some time breaking the loose pieces of banded "flow" rhyolite that lines the hillsides along the trail up Strawberry Butte. With diligent searching, a pocket containing excellent clusters can be found.

Deer Lodge

See map on page 81
Land type: Mountains.
GPS: N. 46.37379 / W. 112.58399
Best season: Spring through fall.
Land manager: USDA Forest Service, Beaverhead–Deerlodge National Forest.
Material: Pyrite, arsenopyrite, chalcopyrite, malachite, quartz crystals, galena, sphalerite.
Tools: Rock hammer.
Vehicle: Any.
Accommodations: Camping and RV parking in Helena National Forest and Beaverhead-Deerlodge National Forest; camping, RV parking, and motels in Deer Lodge.
Special attractions: The Old Montana Prison in Deer Lodge operated for more than one hundred years. Self-guided tours are available; admission includes access to four small museums located on the same grounds.
Finding the site: From Interstate 90 in Deer Lodge, take exit 184 to the frontage road on the east side of the interstate. Drive south on the frontage road for a little less than 1 mile and turn left onto Emery Road. Drive 6.5 miles on Emery Road to a fork, stay right, and continue 0.5 mile farther to tailings piles on the left side of the road and several abandoned mining buildings. Be aware of private property boundaries. At the time of my visit, the right side of the road was not only private but also being mined. As always, respect private property boundaries and remember that there are plenty of old tailings piles on public land in the Zosell Mining district. *DeLorme: Montana Atlas & Gazetteer:* Page 39 B7.

Rockhounding

Mining in the Zosell District began in 1872 when placer gold was discovered. This was followed by the mining of lode deposits in 1887. Mining of gold, silver, copper, and lead continued through the late 1930s.

The rocks that contained the metalliferous deposits are Late Cretaceous volcanic rocks, possibly related to the intrusion of the nearby Boulder batholith. The tailings piles are large and located throughout the district; they produce nice samples of pyrite, arsenopyrite, quartz crystals, and malachite, along with other associated minerals.

Basin

See map on page 81
Land type: Mountains, road cut.
GPS: N. 46.26637 / W. 112.31348
Best season: Any.
Land manager: Jefferson County.
Material: Barite crystals.
Tools: None.
Vehicle: Any.
Accommodations: Camping and RV parking nearby in Helena National Forest and Beaverhead-Deerlodge National Forest; camping, RV parking, and motels in Butte.
Special attractions: The World Museum of Mining and Hell Roaring Gulch just outside of Butte has displays on mining history and a reconstructed 1800s mining town with more than fifty buildings. On the grounds of Montana Tech University in Butte is the Mineral Museum, which has more than 1,500 minerals on display, including what may be the largest gold nugget ever found in Montana.
Finding the site: From Interstate 15 west of Boulder, take Basin exit 156 to the frontage road on the south side of the freeway. Drive 3.7 miles west on the frontage road to a large towering rock outcrop on the left side of the road and park in one of the turnouts. *DeLorme: Montana Atlas & Gazetteer:* Page 39 B8.

Rockhounding

Although there are numerous gold mines in the general vicinity of Basin, some of the more interesting collecting can be done just a few miles west of Basin at Indian Head Rock. At this site large specimens of golden barite crystals can be found. The crystals average about half an inch in length and are so abundant in loose chunks of rock eroding from the cliff that tools aren't even necessary. Gem-quality crystals aren't likely to be found, but the crystals that are here are easy to find and may be several inches long.

Linda Bruner, co-owner of A&L Shoppers Pawn and Rock Shop in Butte, displays the large clusters of golden barite crystals she found at Indian Head Rock.

Flint Creek Hill

See map on page 81

Land type: Mountains, road cut.

GPS: N. 46.22096 / W. 113.28793

Best season: Spring through fall.

Land manager: Montana DOT.

Material: Mudcracks, ripple marks, raindrop imprints.

Tools: Rock hammer.

Vehicle: Any.

Accommodations: Camping and RV parking in Beaverhead-Deerlodge National Forest and on Georgetown Lake; camping, RV parking, and motels in Philipsburg and Anaconda.

Special attractions: The Granite County Museum and Cultural Center in Philipsburg has displays on mining and Montana history, including a sample of a Philipsburg raw ore vein. In Anaconda at the historic district, you can view a 585-foot-tall smokestack—all that remains of Marcus Daly's 19th-century smelter. About 4 miles east of Philipsburg are the remains of the ghost town of Granite, one of the more colorful mining camps of Montana's past.

Finding the site: Philipsburg is located between Glacier National Park and Yellowstone National Park along the Pintlar Scenic Route, a loop connecting to Interstate 90 on both ends. From Philipsburg drive south on Montana Highway 1 for about 10 miles to Flint Creek Hill, a large road cut between Georgetown Lake and the Philipsburg Valley. Towering vertically lifted Precambrian mudstones are on both sides of the road. Even the boulder-size slabs of Precambrian rock placed by the highway department in the turnout to keep cars from plunging into Flint Creek hold specimens of mudcracks that weather away from the rock in nicely split layers. *DeLorme: Montana Atlas & Gazetteer:* Page 38 B3.

Rockhounding

This is by far the best location in Montana to get in touch with Precambrian time. The Precambrian sedimentary layers of the Belt series are at a vertical stance and tower above the highway at a minimum of 30 feet high. In these layers you can see the evidence of a time long since gone on earth. The billion-year-old Precambrian stone contains well-preserved mudcracks and ripple

Mudstone cliffs with mudcracks, ripple marks, and raindrop impressions.

marks and reports of raindrop impressions. This rock started as the sediments of a shallow sea that once covered the land on earth before the atmosphere was rich in oxygen and before animals existed. The imprints in the mudstone technically aren't considered fossils, because they leave no evidence of life. They serve as evidence of a time before life, a time before animal tracks or even burrowing casts from worms. The layers of imprints separate easily, and tools are not required because the road work has flaked off good pieces.

Philipsburg

See map on page 81

Land type: Mountains.

GPS: N. 46.32676 / W.113.26647

Best season: Spring through fall.

Land manager: USDA Forest Service, Beaverhead-Deerlodge National Forest.

Material: Rhodochrosite, galena, silver, sphalerite, pyrite, pyrolusite, arsenopy-rite, tetrahedrite.

Tools: Rock hammer.

Vehicle: Any.

Accommodations: Camping and RV parking in Beaverhead-Deerlodge National Forest; camping, RV parking, and motels in Philipsburg and Anaconda.

Special attractions: The Granite County Museum and Cultural Center in Philipsburg has displays on mining and Montana history, including a sample of a Philipsburg raw ore vein. In Anaconda at the historic district, you can view a 585-foot-tall smokestack—all that remains of Marcus Daly's 19th-century smelter. About 4 miles east of Philipsburg are the remains of the ghost town of Granite, one of the more colorful mining camps of Montana's past.

Finding the site: The city of Philipsburg is located between Glacier National Park and Yellowstone National Park along the Pintlar Scenic Route, a loop connecting to Interstate 90 on both ends. Philipsburg publishes a free newspaper guide for tourists that is available throughout the town. The *Philipsburg Territory* is updated every year and includes a historic driving tour of Philipsburg and the mining district just outside of town. The tour is printed in the paper along with a nice map. Each mine is identified on the map, and the land manager is noted. Be aware that public and private land is intermingled throughout the district. It is a good idea to pick up one of these maps since many of the mines are on private property. Always be on the lookout for any indication of private property on the dozens of abandoned mines east of Philipsburg. It is also recommended that you inquire locally on land status as well. For a copy of the *Philipsburg Territory* or more information, contact Philipsburg Chamber of Commerce, P.O. Box 661, Philipsburg, MT 59858; (406) 859–3388; www.philipsburgmt.com. *DeLorme: Montana Atlas & Gazetteer:* Page 38 B3.

Rockhounding

The Philipsburg District was a relatively important one during the early 1900s, after the discovery of gold and silver near Georgetown to the south. Several deposits have been exploited in the district, and the presence of a smelter indicates an extensive source of ore.

The occurrence of such minerals as pyrolusite, psilomelane, and rhodochrosite attest to the fact that manganese was one of the major elements being searched for here. Fragments of the host rock are common on the dumps and tailing, and cavities or vugs in this rock may reveal outstanding tiny crystals of manganese minerals. The possibility of obtaining some interesting micromount specimens in this area is good.

The best luck in collecting good samples from this large district may be to use the *Philipsburg Territory* map of the mining district and spend a day driving throughout the district and exploring some of the dozens of tailings piles in the area. There are a lot of interesting minerals to be found here, so don't forget to bring a rock and mineral identification guide—who knows what you may find!

Gem Mountain

See map on page 81
Land type: Mountains.
GPS: N. 46.24707 / W. 113.59235
Best season: Generally open seven days a week mid–May to mid–October.
Land manager: Private, Gem Mountain Cooney's Sapphire Village, Inc.
Material: Sapphire.
Tools: Supplied by business.
Vehicle: Any.
Accommodations: Camping and RV parking nearby in Beaverhead-Deerlodge National Forest; camping, RV parking, and motels in Philipsburg and Anaconda.
Special attractions: The Granite County Museum and Cultural Center in Philipsburg has displays on mining and Montana history, including a sample of a Philipsburg raw ore vein. In Anaconda at the historic district, you can view a 585-foot-tall smokestack—all that remains of Marcus Daly's 19th-century smelter. About 4 miles east of Philipsburg are the remains of the ghost town of Granite, one of the more colorful mining camps of Montana's past.
Finding the site: Philipsburg is located between Glacier National Park and Yellowstone National Park along the Pintlar Scenic Route, a loop connecting to Interstate 90 on both ends. From Philipsburg drive south on Montana Highway 1 for 6 miles to the junction of Montana Highway 38. Take MT 38 (Skalkaho Pass) 16 miles west to Gem Mountain. There are signs for Gem Mountain posted on MT 38, and a large welcome sign is at the driveway to the mine. *DeLorme: Montana Atlas & Gazetteer:* Page 38 B2.

Rockhounding

The sapphire deposits at the Gem Mountain locality are somewhat similar to those in the Helena area, but they were not deposited by a stream the size of the Missouri River. These deposits include bench and terrace gravels that were laid down by mountain streams.

Mining began at Gem Mountain in the early 1890s, and since that time it has produced more than 180 million carats in sapphires. It has been reported that the sapphires found at Gem Mountain exhibit a wider range of colors (from yellow to pink and lavender) than those found at other localities in the state.

The office at Gem Mountain.

There is no fee for digging at Gem Mountain, but customers may purchase buckets of pre-dug gravel from the mine for screening on the premises. You're not allowed to bring your own tools, but everything you need is provided on-site, along with a staff to help you screen your purchases. Bags of prescreened "pay dirt" concentrate are available for purchase, and bags can be shipped worldwide if customers are interested in looking for sapphires in the comfort of their own home. The owners of the mine also provide a faceting service and trained jewelers are on-site, along with a gift shop with a good selection of fine jewelry for purchase.

For more information or to order gravel through the mail, contact Gem Mountain Cooney's Sapphire Village, Inc., P.O. Box 148, 3835 Skalkaho Road, Philipsburg, MT 59858; (866) 459–GEMS; www.gemmtn.com.

Homestake Pass

See map on page 81
Land type: Mountains.
GPS: N. 45.92651 / W. 112.40120
Best season: Spring through fall.
Land manager: USDA Forest Service, Beaverhead–Deerlodge National Forest.
Material: Quartz, schorl, feldspar, garnet.
Tools: Rock hammer, sledge hammer, and chisel.
Vehicle: Any.
Accommodations: Camping and RV parking nearby in Beaverhead–Deerlodge National Forest at Delmoe Lake; camping, RV parking, and motels in Butte.
Special attractions: The World Museum of Mining and Hell Roaring Gulch just outside of Butte has displays on mining history and a reconstructed 1800s mining town with more than fifty buildings. On the grounds of Montana Tech University in Butte is the Mineral Museum, which has more than 1,500 minerals on display, including what may be the largest gold nugget ever found in Montana.
Finding the site: Pegmatites and loose large shaft crystals can be found throughout the Homestake Pass and Delmoe Lake area. To collect the best specimens, use a map as a guide and spend a day or two exploring the area. The site mentioned is the closest known locality to the highway.

From Interstate 90 southeast of Butte, take exit 233 (Homestake). Go to the frontage road on the north side of the interstate and follow the road that leads toward Delmoe Lake (Forest Road 222). Drive 0.8 mile and turn left onto an unmarked road. Drive just a few hundred feet and park at a small turnout on the right side of the road. Look for the hill on the right side of the road that shows evidence of digging—a good place to start the search. *DeLorme: Montana Atlas & Gazetteer: Page 39 D7.*

Rockhounding

In the pegmatites of the Boulder batholith, nice specimens of quartz, schorl, feldspar, and garnet can be found. Pegmatites are an especially coarse-grained igneous rock (in this case the composition is granite) that often contain large amounts of minerals due to the rock's formation during the final and most hydrous stage of the cooling of the batholith when the magma is crystallized. The best way to search for a nice sample of Homestake pegmatites is to look for an area where others have been digging and start digging. Splitting the rock in and around the worked areas can also prove to be productive, but it all depends on the luck of the area you work. There is no shortage of virgin places to go prospecting either.

The feldspar and garnet are not defined crystals, but the chunks can be quite large. There have been reports of schorl crystals several inches long, although none have been personally observed. It seems that most rockhounds in the area are searching for the quality quartz crystals to be found. If any crystals are found by digging pegmatites, they will be difficult to remove from the rock without damaging them.

There is another way to search for complete quartz crystals, however. Several pegmatites with quartz crystal-lined vugs outcrop along the north shore of Delmoe Lake. The action of the waves upon these rocks apparently breaks the crystals loose. They can periodically be found in the sand and gravel along the shore during periods of low water. The crystals are not large, but they often possess nice terminations.

Rader Creek

See map on page 81
Land type: Mountains.
GPS: N. 45.88747 / W. 112.36748
Best season: Late spring through fall.
Land manager: USDA Forest Service, Beaverhead-Deerlodge National Forest.
Material: Quartz, mica, feldspar.
Tools: Rock hammer, sledge hammer, pick, and shovel.
Vehicle: High clearance.
Accommodations: Camping and RV parking in Beaverhead-Deerlodge National Forest; camping, RV parking, and motels in Butte.
Special attractions: The World Museum of Mining and Hell Roaring Gulch just outside of Butte has displays on mining history and a reconstructed 1800s mining town with more than fifty buildings. On the grounds of Montana Tech University in Butte is the Mineral Museum, which has more than 1,500 minerals on display, including what may be the largest gold nugget ever found in Montana.
Finding the site: From the junction of Montana Highways 2/41 southwest of Butte, take MT 2 4.6 miles west of the junction and turn right onto Toll Mountain Road. Drive 1.3 miles and turn right onto Upper Rader Creek Road (Forest Road 240). Continue for 2 miles from the intersection with Whiskey Gulch Road, staying on Upper Rader Creek Road, to a large area on the right in a small clearing where people have been digging. (These directions are merely a starting point for collecting in the Toll Mountain area.) *DeLorme: Montana Atlas & Gazetteer:* Page 39 D7.

Rockhounding

The rocks in the Rader Creek area between Toll Mountain and Interstate 90 have produced some of the finest smoky quartz crystals found in the state. The quality crystals are several inches long in size and can sometimes be found in a very dark smoky shade. The crystals occur in pockets (miarolitic cavities) throughout the granite-like rocks of the Boulder batholith, which is exposed in this region. The area is a popular one for local rockhounds, and any material lying free on the surface has more than likely been picked up. Digging is perhaps the only way one will obtain good smoky quartz here, but it would

Diligently digging in search of a crystal pocket.

help to seek additional information from local rock shops or rock-club members as to its location and availability.

The easiest way to search for crystals in the Rader Creek area is to look for areas where other people have been digging. Good specimens tend to come from locations of quartz that is already exposed. Also be on the lookout for large pegmatites in the country rock and evidence that someone else has been collecting out of the same location. Oftentimes, a pick and shovel are necessary to break anything free and, due to the size of most of the pits, hand tools larger than a rock hammer are recommended.

Calvert Loop

See map on page 81

Land type: Mountains.

GPS: N. 45.84757 / W. 113.15343

Best season: Summer.

Land manager: USDA Forest Service, Beaverhead Deerlodge National Forest.

Material: Epidote crystals, garnet, calcite crystals, calcite, aquamarine, scheelite.

Tools: Rock hammer, sledge hammer, chisels.

Vehicle: High clearance.

Accommodations: Camping and RV parking in Beaverhead-Deerlodge National Forest; camping, RV parking, and motels in Anaconda and Butte.

Special attractions: The World Museum of Mining and Hell Roaring Gulch just outside of Butte has displays on mining history and a reconstructed 1800s mining town with more than fifty buildings. On the grounds of Montana Tech University in Butte is the Mineral Museum, which has more than 1,500 minerals on display, including what may be the largest gold nugget ever found in Montana.

Finding the site: From the junction of Montana Highways 43/569 between Wisdom and Wise River, drive east on MT 43 for 3.5 miles and turn right into the Dickey Bridge Recreation Area. Drive past the camping and picnicking areas and stay on Forest Road 1213 for 4.2 miles to a fork in the road—this is the Calvert Loop. Stay right at the fork and drive 2.5 miles to the site. There

Epidote crystal.

Rockhound Kellyn Wagner displays her much-needed sledge hammer in front of the Calvert mine.

is a large pit filled with water on the right side and a field of dump piles on the left. Park at the farthest side and explore. *DeLorme: Montana Atlas & Gazetteer:* Page 38 D3.

Rockhounding

The Calvert tungsten mine, and the small Wise River Mining district itself, is a good example of the search for nonprecious metals after the slowdown of the Montana gold and silver rush. Mining at Calvert took place periodically on a relatively small-scale basis from 1956 to 1966. Massive dumps are spread out as testament to the 113,000 tons of ore that were mined during the ten-year run.

There is a tremendous amount of interesting rocks and minerals to be found at the Calvert mine. The mine itself looks more like an old quarry. Today a large pool of water covers the deep pit where part of a mountain once stood.

Acres of old dump piles offer a great opportunity for rockhounds to collect large epidote crystals and large but fractured garnet samples. The pieces are all loose. Calcite crystals are found throughout the dumps and the area. With luck, large chunks of white calcite can be diligently split to reveal the most precious gem of them all—aquamarine, some of gem quality and up to several inches long. On one occasion the Butte Mineral and Gem Club was at the site with some nice pieces of aquamarine they discovered. Splitting the calcite is not easy, and a sledge hammer along with a nice set of chisels is needed. The prospect of finding precious gems might make the extra work worth a try.

Quartz Hill

See map on page 105
Land type: Mountains.
GPS: N. 45.71737 / W. 112.91872
Best season: Late spring through fall.
Land manager: USDA Forest Service, Beaverhead-Deerlodge National Forest.
Material: Quartz, quartz crystals.
Tools: Rock hammer, sledge, chisel.
Vehicle: High clearance.
Accommodations: Camping and RV parking in Beaverhead-Deerlodge National Forest; camping, RV parking, and motels in Butte.
Special attractions: None.
Finding the site: From Montana Highway 43 at the west side of Dewey, turn south onto Quartz Hill Road. Drive about 5.5 miles to the old mine tailings and an additional 1.5 miles to Quartz Hill. *DeLorme: Montana Atlas & Gazetteer: Page 25 A5.*

Rockhounding

The Quartz Hill mining district produced significant amounts of ore from silver-bearing quartz veins spliced through the Paleozoic limestone. Along with silver, the primary mineral mined, the district also produced gold, copper, lead, and zinc. The tailings piles are the only thing left to tell the tale of this once-bustling area, and even they have little to say. Suitably, quartz and quality quartz crystals are the most abundant specimens to be collected here. The piles produce some small samples of crystals, but for the best material and the largest crystals you'll have better luck using a forest map to explore several small forest roads that lead up to the top of Quartz Hill. Nice samples of quartz and crystals seem to be along these roads if you chase the vertical veins up the hills.

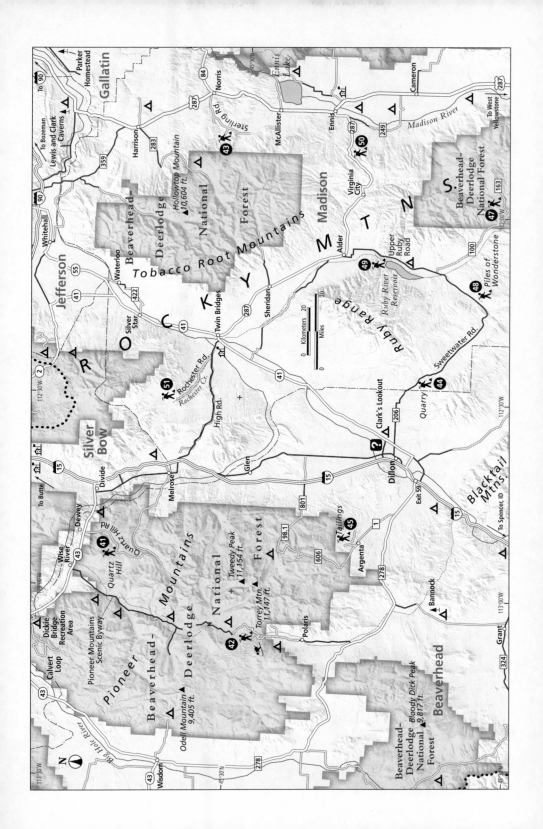

Crystal Park

See map on page 105
Land type: Mountains.
GPS: N. 48.48702 / W. 113.10024
Best season: Crystal Park is generally open May 15—October 15.
Land manager: USDA Forest Service, Beaverhead-Deerlodge National Forest.
Material: Quartz crystals.
Tools: Shovel, ¼-inch screen.
Vehicle: Any.
Accommodations: Camping and RV parking nearby in Beaverhead-Deerlodge National Forest; camping, RV parking, and motels in Dillon.
Special attractions: Bannack State Historic Park is about 25 miles south of Crystal Park on the way to Dillon. The preserved ghost town of Bannack was Montana's first territorial capital and one of the original gold-rush towns. Today there is a visitor center and several historic buildings, as well as many tailings piles and mining equipment.
Finding the site: From just west of Wise River, drive south on the Pioneer Mountains Scenic Byway for 27 miles and turn right at the sign for Crystal Park. Park at the parking area and don't forget to pay your daily entrance fee (at the time of my visit it was $5.00 per vehicle). If you're coming from the southern entrance, take Montana Highway 278 to the Scenic Byway, drive north 17.8 miles, and turn left at the sign for the park. *DeLorme: Montana Atlas & Gazetteer:* Page 24 B4.

Rockhounding

Peaceful meadows, beautiful forests, and majestic peaks provide an ideal setting for Crystal Park, where rockhounds may dig for quartz crystals. The crystals were formed as a result of the Pioneer batholith, intruding up around sixty-eight million years ago and creating the Pioneer Mountains. After a lengthy time of cooling, the crystals formed within small pockets in the granite. Thanks to millions of years of weathering and erosion, the crystals can be found already weathered from the granite and settled loose in the soil.

The many pits occurring on the slopes suggest this locality is popular with rockhounds. The area is extensive; at the time I visited more than thirty acres were open to the public to explore. Crystals of clear and smoky quartz and

Some crystals from a day at Crystal Park.

even amethyst can be found with conscientious effort. They are found only by digging in the hillside, and this means work.

The serious collector should plan to bring a shovel, pick, and some assorted mesh screens to sift the dirt and loose rock. The crystals are scattered throughout the soil in the park, and a good place to start collecting is in a pit that was already worked. Sometimes reddish layers in the soils indicate a more-condensed collection of crystals. Extremely nice crystals have been taken from this locality, some several inches long, but quality comes with time and patience. Crystal Park is open for day use, and at least one full day of exploration is recommended.

Several nearby campgrounds offer adequate facilities for those who wish to remain for a spell and enjoy some very beautiful scenery, as well as a "good dig." Crystal Park itself has picnic tables and well-maintained restroom facilities. This is a wonderful site for children; nice paths lead to the digging areas, playing in dirt is always fun, and no one seems to go home empty-handed.

It is hoped that those who visit Crystal Park will not abuse their privilege. Though the area is relatively clean, there is evidence of littering and vandalism, and some trees have fallen over due to careless digging under the roots in violation of posted rules. A concerted effort on the part of those who visit here will ensure a continuing source of fine crystal specimens for future generations.

For a brochure on Crystal Park along with rules and digging tips, write to Dillon Ranger District Office, Beaverhead-Deerlodge National Forest, 420 Barrett Street, Dillon, MT 59725; (406) 683–3900.

Norris

See map on page 105
Land type: Valley.
GPS: N. 45.53937 / W. 111.78531
Best season: Spring through fall.
Land manager: Bureau of Land Management (BLM).
Material: Quartz crystals.
Tools: Rock hammer.
Vehicle: Any.
Accommodations: Camping, RV parking, and motels in Three Forks and Ennis.
Special attractions: The reconstructed gold-rush town of Virginia City has more than eighty buildings and a rich, wild history. The town hosts a small mining museum, a brewery, and a tourist "railroad" that runs a small gas-powered

Rockhound Jessie Dart splits a chunk of quartz in search of the perfect crystal.

locomotive to move passengers between Virginia City and the living ghost town of Nevada City. Both towns are now popular tourist attractions during the summer months, and many of the original buisnesses have been restored. Numerous curios and gift shops provide souvenirs and some occasionally offer lessons in the art of panning gold.

Finding the site: From Norris at the junction of U.S. Highway 287 and Montana Highway 84, turn west onto Sterling Road, the dirt road directly across from MT 84. Take the road for 3.2 miles, turn left, continue for an additional 2.7 miles, and turn right onto a small dirt road. At this point you should be able to see tailings piles; a good site is 0.2 mile down this road. Park near the pits and look around at the loose rock on the ground for quartz crystals. *DeLorme: Montana Atlas & Gazetteer:* Page 26 B3.

Rockhounding

The quartz crystals found in the tailings piles of the mining district just west of Norris are a result of the quartz monazite intrusion that runs throughout the area. In the 1860s gold was discovered and mined on a small scale through several methods until the early 1900s. Today, small tailings piles and shallow pits spread across the barren valley are all that remain.

Nice but very small quartz crystals can be found easily in the abundant piles of white quartz tailings that cover the area. The translucent crystals are difficult to break free from the parent milky quartz, and the best samples to keep for collections are clusters of quartz crystals still attached to the rock. Splitting rocks that show potential may reveal spectacular pockets.

Dillon

See map on page 112
Land type: High hills.
GPS: N. 45.17422 / W. 112.42529
Best season: Spring through fall.
Land manager: Beaverhead County.
Material: Soapstone.
Tools: None.
Vehicle: Any.
Accommodations: Camping, RV parking, and motels in Dillon.
Special attractions: Bannack State Historic Park is about 20 miles west of Dillon on Montana Highway 278. The preserved ghost town of Bannack was Montana's first territorial capital and one of the original gold-rush towns. Today there is a visitor center and several historic buildings, as well as many tailings piles and mining equipment.
Finding the site: From Dillon go about 1 mile north of town and turn right onto Sweetwater Road (County Road 206). Drive 11.5 miles to the quarry on

A piece of raw soapstone from Sweetwater Road and a heart-shaped box carved from the Sweetwater Road rock.

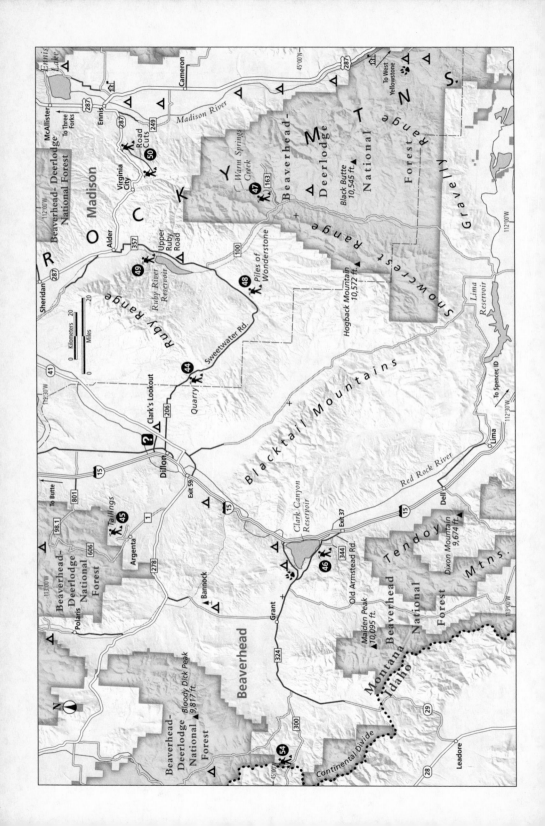

the right. Make sure to park away from the quarry and not to disturb quarry traffic or block the quarry's driveway. The Regal Mine is owned by Barretts Minerals, Inc, and is on private land. It is uncertain if the land manager would grant permission to collect on dumps within the private property. *DeLorme: Montana Atlas & Gazetteer:* Page 25 D7.

Rockhounding

The Regal Mine southeast of Dillon has mined talc from the open pit since 1972. The quarry is private and was active at the time I visited. This shouldn't discourage the collector, however. There is enough interesting material to be found dumped alongside the road to keep generations of rockhounds happy. Nice specimens of soapstone, up to several pounds heavy, can be found discarded on the side of the road and in the tailings piles that line the road. The green material is quite soft and can be carved easily. When polished, it takes on the look of quality jade.

Argenta Mine Tailings

See map on page 112

Land type: High hills.

GPS: N. 45.28766 / W.112.86794

Best season: Spring through fall.

Land manager: Bureau of Land Management (BLM).

Material: Calcite crystals, pyrite, malachite, jasperoid, chert.

Tools: Rock hammer.

Vehicle: Any.

Accommodations: Camping, RV parking, and motels in Dillon.

Special attractions: Bannack State Historic Park is about 20 miles west of Dillon on Montana Highway 278. The preserved ghost town of Bannack was Montana's first territorial capital and one of the original gold-rush towns. Today there is a visitor center and several historic buildings, as well as many tailings piles and mining equipment.

Finding the site: From Interstate 15 in Dillon, drive south for 2.5 miles and take exit 59 to Montana Highway 278. Drive 6.7 miles and turn right onto the road leading to Argenta. Continue 5.6 miles, turn right onto a dirt road, and drive 0.1 mile to another dirt road and turn right. If you're in a car without high-clearance four-wheel drive, park here and walk up the road 0.2 mile to the tailings piles; otherwise, drive up the road and park at one of the many mines. It should be noted that there are open shaft mines at the top of the hill, and extreme caution should be exercised. Children visiting the site is not recommended. *DeLorme: Montana Atlas & Gazetteer:* Page 25 C5.

Rockhounding

Mining did not begin in the Argenta area until Bannack had been fully explored. Unlike the Bannack region, however, silver and lead were the more important metals produced from the Argenta District. Also, very little placer activity took place, so the vast majority of the production came from the lode deposits.

The type of rock exposed in the Argenta area is similar to that exposed at Bannack. It is sedimentary, but in general much older. Most of the ore bodies occur in these rocks where mineral-bearing solutions were apparently injected either into fissures or along bedding planes, but some contact metamorphism deposits do exist within the district.

Argenta minerals.

Watch out for pits that may show signs of activity, and be on the lookout for posted claims before collecting mineral specimens here. Much of the ore mined has been extremely oxidized, but certain sulfides as well as various copper carbonates occur on the dumps and tailings of the old mines.

Clark Canyon Reservoir

See map on page 112
Land type: Hills.
GPS: N. 44.93583 / W. 112.87526
Best season: Late spring through fall.
Land manager: Bureau of Land Management (BLM).
Material: Fossils, gypsum.
Tools: Rock hammer.
Vehicle: High clearance.
Accommodations: Camping and RV parking on Clark Canyon Reservoir; camping, RV parking, and motels in Dillon.
Special attractions: None.
Finding the site: From Interstate 15 in Dillon, drive south about 30 miles and take exit 37 to the west side of the interstate. Drive north on Old Armstead Road for 2.8 miles and stay left at the fork. At 0.3 mile after the fork, turn left onto County Road 344 and drive 0.7 mile to two-track dirt roads on the right. Both of these roads enter the BLM land. A high-clearance vehicle is recommended for these roads; otherwise, walk 1 mile down the BLM road to various limestone outcrops. *DeLorme: Montana Atlas & Gazetteer:* Page 20 A4.

Rockhounding

This is one of those park-and-explore sites since the Madison limestone is tricky to pinpoint in one location. Outcrops of the 350-million-year-old sedimentary rock resulted from a time when the area was covered with a shallow sea. The dark gray limestone is scattered around the lake area, but collecting a sample containing fossils can be challenging. Mediocre specimens of horn corals and brachiopods were found in loose pieces around the area, particularly in the washes. Of course, their source was undetermined. With a BLM map and a nice day to walk the area, good specimens might be found. The gypsum seems to be more abundant at the entrance to the BLM hills, but it can be found in the washes as well. For a quality specimen, or even a small crystal, you'll have to hike up to their source on the ridges. A sunny day is helpful in locating them.

Warm Springs Creek

See map on page 112
Land type: High hills.
GPS: N. 45.04305 / W. 112.93784
Best season: Late spring–fall.
Land manager: USDA Forest Service, Beaverhead–Deerlodge National Forest.
Material: Fossils.
Tools: Rock hammer.
Vehicle: Any.
Accommodations: Camping and RV parking on the Ruby Reservoir; camping, RV parking, and motels in Virginia City and Ennis.
Special attractions: The reconstructed gold-rush town of Virginia City has more than eighty buildings and a rich, wild history. The town hosts a small mining museum, a brewery, and a tourist "railroad" that runs a small gas–powered

Nevada City is a popular tourist attraction.

A chunk of limestone containing snail fossils.

locomotive to move passengers between Virginia City and the living ghost town of Nevada City. Both towns are now popular tourist attractions during the summer months, and many of the original businesses have been restored. Numerous curios and gift shops provide souvenirs and some occasionally offer lessons in the art of panning gold.

Finding the site: From Alder, just west of Virginia City, drive south on Upper Ruby Road (Montana Highway 357/ Forest Road 100) for 26.7 miles and turn left onto Forest Road 163. Drive 2.2 miles to the site and park on the left. Warm Springs Creek is on the north side, and large weathering limestone hills are along the creek on the left. *DeLorme: Montana Atlas & Gazetteer:* Page 26 D2.

Rockhounding

There is a multitude of interesting fossils to be found here. At the time I visited, abundant fossils of Jurassic oysters and snails were easily collected by searching the area between the limestone cliffs and Warm Springs Creek. Enough good specimens are already exposed on the ground that there is no need to split for fossils. It is a good thing too, since the limestone is not easy to break.

The immense tailings between Virginia City and Alder are mute testimony to the intense search for gold that once occurred in this area. Huge dredges chewed into the stream gravel, spit out the unwanted rock, and kept the gold within. Gold still lies hidden beneath the rugged terrain, but even with present gold prices, costs keep the number of operating mines to a minimum.

Sweetwater Road

See map on page 112
Land type: High hills.
GPS: N. 45.07976 / W. 112.21957
Best season: Any.
Land manager: Bureau of Land Management (BLM).
Material: Wonderstone.
Tools: Rock hammer, sledge hammer.
Vehicle: Any.
Accommodations: Camping and RV parking on the Ruby Reservoir; camping, RV parking, and motels in Virginia City, Ennis, and Dillon.
Special attractions: The reconstructed gold-rush town of Virginia City has more than eighty buildings and a rich, wild history. The town hosts a small mining museum, a brewery, and a tourist "railroad" that runs a small gas-powered locomotive to move passengers between Virginia City and the living ghost town of Nevada City. Both towns are now popular tourist attractions during the summer months, and many of the original businesses have been restored. Numerous curios and gift shops provide souvenirs and some occasionally offer lessons in the art of panning gold.
Finding the site: From Alder drive south on Upper Ruby Road (Montana Highway 357) for 13.5 miles and turn right on the road leading to Dillon (Sweetwater Road). Drive 6.8 miles past the second cattle guard to large hills of eroding volcanic rock. Park on either side of the road in the small turnouts. *DeLorme: Montana Atlas & Gazetteer:* Page 26 C1.

Rockhounding

The volcanic rock often referred to as "Montana Wonderstone" is a very attractive rock with bands of varying widths. Shades of yellow, brown, and red make up the colors of the bands. Technically described as silicified interbedded tuff, this material is suitable for decorative bookends and similar products. The rock does not take a high polish and therefore does not fashion well into cabochons, although some of good quality can be found. The rock was quarried as a source of terrazzo, but apparently the operation was abandoned after a short period of time. The quarry containing the rock is located on private land and trespassing is not allowed.

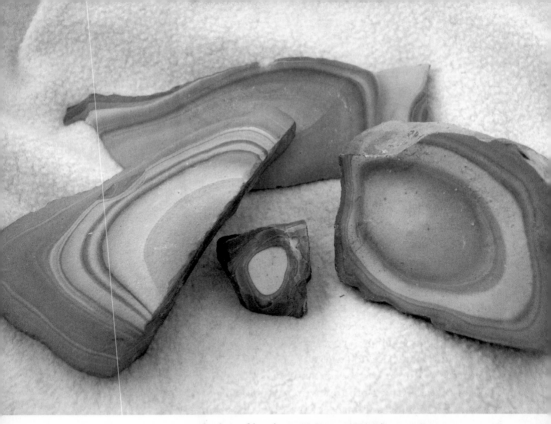

The amazing patterns and colors of bands on "Montana Wonderstone."

A sympathetic soul or possibly a disgruntled landowner dumped several heaping piles of the material alongside the left side of the road. The reasoning behind this is unknown, but the piles are open to rockhounds. Thankfully so, and hopefully the piles will remain; otherwise, it could be difficult to obtain large pieces. The rock looks like an ordinary dark brown to red volcanic tuff on the exposed layers; the tough tuff must be split to expose the patterned treasures. Splitting the layers requires a good deal of safety precautions. Protective eyewear, long sleeves, pants, and gloves are recommended as the chips of the rock have a tendency to pierce forcefully into the rockhound as they're split. If the piles should for some reason no longer be available along the road, small samples can be collected from splitting the tuff eroding from the road cuts at the same location. The smaller pieces do tend to have more detailed and colorful patterns.

Ruby Reservoir

See map on page 112
Land type: Lakeshore.
GPS: N. 45.23624 / W. 112.10707
Best season: Early spring or late summer through fall.
Land manager: Montana Fish, Wildlife and Parks.
Material: Garnets.
Tools: Screen (preferably about ⅛-inch mesh), shovel, small container for garnets.
Vehicle: Any.
Accommodations: Camping and RV parking on-site; camping, RV parking, and motels in Virginia City and Ennis.
Special attractions: The reconstructed gold-rush town of Virginia City has more than eighty buildings and a rich, wild history. The town hosts a small mining museum, a brewery, and a tourist "railroad" that runs a small gas-powered locomotive to move passengers between Virginia City and the living ghost town of Nevada City. Both towns are now popular tourist attractions during the summer months, and many of the original businesses have been restored. Numerous curios and gift shops provide souvenirs and some occasionally offer lessons in the art of panning gold.
Finding the site: The only specification for selecting an area to collect garnets is to be on the reservoir at a time when the water level is low. One recommendation for a nice place to park and search the shore can be reached from Alder by driving south on Upper Ruby Road (Montana Highway 357) for 7.5 miles and then turning right onto any of the several dirt roads that lead to the shore of the Ruby Reservoir. *DeLorme: Montana Atlas & Gazetteer: Page 25 D8.*

Rockhounding

Perhaps the most impressive material to be found in the Virginia City area is garnet, which abounds in some of the dark metamorphic rock in the region. And no rock hammer will be required for the best way to search for these gems.

Gem-quality almandine garnets, suitable for faceting, are frequently found by screening the terrace and stream deposits in the valleys of the Ruby River and Sweetwater Creek. An ideal place to search for them is along the shores of the Ruby Reservoir above Ruby Dam. The garnets vary in size, although most

Jessie Dart walks the gravel terraces along the Ruby Reservoir on a late summer's afternoon searching for garnets.

are very small. A series of screen meshes would best separate the gravel from the fine sand, and the small gravel can be washed in the reservoir. The distinctive deep-red color of the garnets distinguishes them from the other rock particles. Garnets can also be found without screens during time of low water by walking the edges of each small terrace on the lakeshore and studying gravel deposits washed to the shore as the water level dropped. On a sunny day the garnets will shine in their deep red translucency. This method, of course, requires a bit more time than screening.

Obviously, not all the garnets found will be of gem quality, but time and practice should reward collectors with good specimens. The best time to search for garnets along the shore of the reservoir is during periods of low water. Early spring, before the streams begin carrying meltwater from the mountains, or during late summer and fall when the streams are running very low would be ideal. At no time, however, should one dig in the road cuts along the east shore

Ruby Reservoir garnets.

of the reservoir. The road cuts may exhibit some rocks containing a high percentage of garnet associated with the black mineral hornblende, but the highway department has placed warning signs against rockhounding within this area. Anyone who disobeys the warning will be prosecuted.

Virginia City

See map on page 112
Land type: High hills, road cut.
GPS: N. 45.30976 / W. 111.85530
Best season: Spring through fall.
Land manager: Montana DOT.
Material: Garnet, quartz, feldspar.
Tools: Rock hammer.
Vehicle: Any.
Accommodations: Camping and RV parking on the Ruby Reservoir; camping, RV parking, and motels in Virginia City and Ennis.
Special attractions: The reconstructed gold-rush town of Virginia City has more than eighty buildings and a rich, wild history. The town hosts a small

Searching the road cut between Virginia City and Ennis.

mining museum, a brewery, and a tourist "railroad" that runs a small gas-powered locomotive to move passengers between Virginia City and the living ghost town of Nevada City. Both towns are now popular tourist attractions during the summer months, and many of the original businesses have been restored. Numerous curios and gift shops provide souvenirs and some occasionally offer lessons in the art of panning gold.

Finding the site: Several road cuts along Montana Highway 287 between Virginia City and Ennis produce quartz, garnet, and feldspar pegmatites. Watch for the volcanic rocks and pink, red, and black layers. Remember to park off the highway in a large turnout or on a cross road. As with all road cuts, use caution around traffic. *DeLorme: Montana Atlas & Gazetteer: Page 26 C2.*

Rockhounding

Volcanic rocks exposed in road cuts along Montana Highway 287 between Virginia City and Ennis contain vugs and cavities lined with zeolite and carbonate minerals. Adjacent to them are dark-colored, banded metamorphic rocks containing red almandine garnet. The metamorphic rocks are frequently cut by pegmatite dikes that contain white, glassy, and sometimes rose-colored quartz, as well as fine cleavable masses of pink microcline feldspar. The view toward the east here is also very spectacular, revealing the rugged glaciated peaks of the Madison Range.

Rochester Mining District

See map on page 128
Land type: Mountains.
GPS: N. 45.62378 / W. 112.50127
Best season: Spring through fall.
Land manager: Bureau of Land Management (BLM).
Material: Quartz, pyrite, mica, galena, sphalerite, malachite, azurite, chrysocolla, arsenopyrite.
Tools: Rock hammer.
Vehicle: Any.
Accommodations: Camping, RV parking, and motels in Dillon.
Special attractions: None.
Finding the site: From Twin Bridges drive south on Montana Highway 41 for about 1 mile and turn right onto the road leading to the High Road Fishing Access. Drive 2 miles, turn right onto Rochester Road, and stay on the main road following the creek bed for about 8 miles to the Rochester Mining district. There are numerous mines and abandoned tailings piles. It should be noted that at the time of my visit the first mine showed signs of activity, so be aware of active mining claims and their postings. Most claims are posted with small metal signs on short posts near the piles. There are plenty of abandoned tailings piles to dig through without poking around the claims. *DeLorme: Montana Atlas & Gazetteer:* Page 25 A7.

Rockhounding

The mineral deposits of the Rochester District were exploited mainly for gold and silver. The ore deposits occur primarily as veins in metamorphic rock, as contact deposits along the margins of Paleozoic limestones and igneous intrusions, or as replacement deposits in limestone. Mining of these deposits began on a small scale during the late 1800s when placer operations were attempted but proved to be too difficult to work due to the lack of water. The early exploration was quickly followed by underground mining, and during the productive years the district produced a fair quantity of high-grade ore.

The dumps and tailings, although deeply weathered, have been reported to produce, with careful searching, specimens of pyrite, galena, sphalerite,

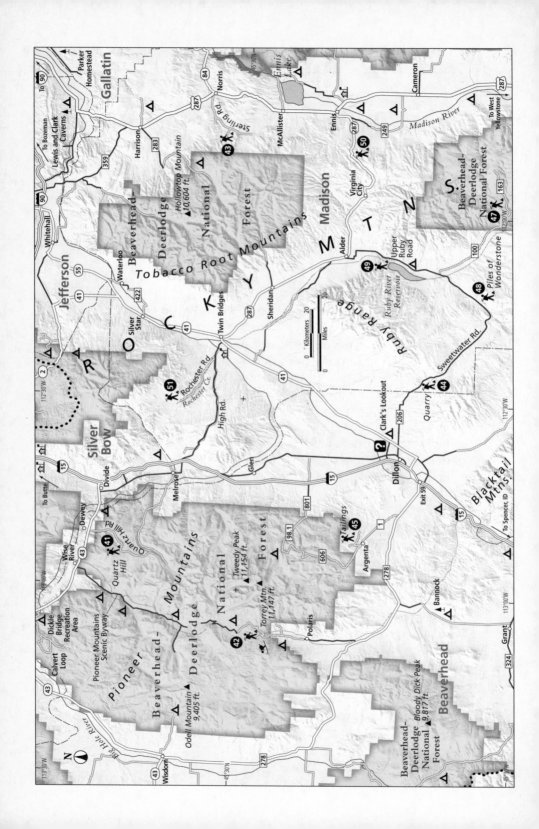

malachite, azurite, chrysocolla, arsenopyrite, quartz, and very rare specimens of cerussite, vanadinite, pyromorphite, and anglesite. At the time of my visit, nice samples of only pyrite, quartz, mica, and azurite were found, and the term "deeply" weathered can again be emphasized. If there was ever a place to view the harsh climate of Montana's rugged winters, it would be most easily described on a mineral specimen from the Rochester District.

Benton Creek

See map on page 131

Land type: Mountains, stream.

GPS: Large and sporadic area.

Best season: Summer through fall.

Land manager: USDA Forest Service, Helena National Forest.

Material: Gold, sapphires.

Tools: Gold pan, trowel, fleck flask.

Vehicle: Any.

Accommodations: Camping and RV parking in Helena National Forest; camping, RV parking, and motels in White Sulphur Springs.

Special attractions: None.

Finding the site: From the junction of U.S. Highway 89 and Montana Highway 360 in White Sulphur Springs, drive 22 miles northwest on MT 360. Turn left at the fork onto Benton Creek Road, which enters Helena National Forest after about 5 miles. Gold pan in the area along Benton Creek.

One site can be reached by driving 2.8 miles after the national forest entrance to a fork in the road. Park on the left and pan the creek, but stay on the east side of the road—the property on the right side of the road is claimed. *DeLorme: Montana Atlas & Gazetteer:* Page 56 D4.

Rockhounding

Gold deposits in the Big Belt Mountains occur as a result of fissures in the ancient Belt rock that have been eroded and deposited once again.

Not only can you pan for gold here, but if you pay close attention to the gravels you're working, you may also find sapphires, mostly fractured but some with good color. Digging for panning material is productive in the area near the creek, and even from the creek itself.

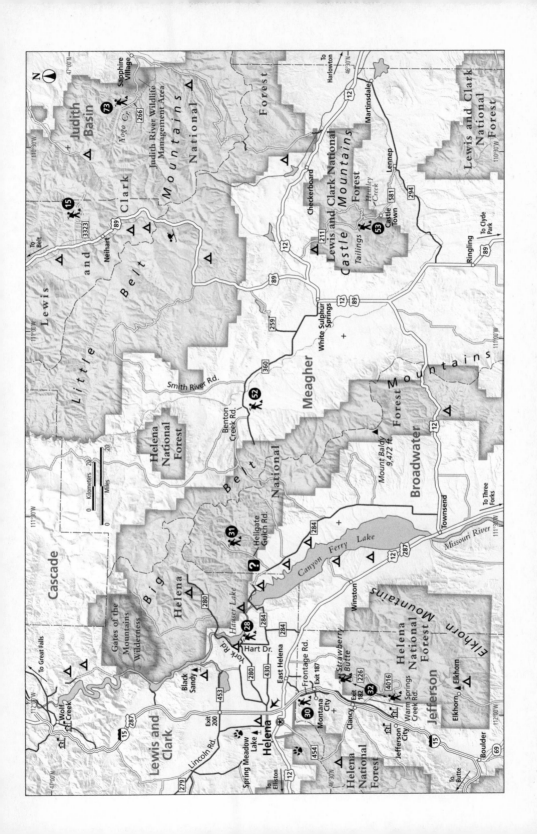

Castle Town

See map on page 131
Land type: Mountains.
GPS: N. 46.48265 / W. 110.69010
Best season: Late spring through fall.
Land manager: USDA Forest Service, Lewis and Clark National Forest.
Material: Malachite, chrysocolla, chalcopyrite.
Tools: Rock hammer.
Vehicle: Any.
Accommodations: Camping and RV parking in Lewis and Clark National Forest; camping, RV parking, and motels in White Sulphur Springs.
Special attractions: None.

A specimen of copper minerals from Castle Town.

Finding the site: It is easiest to approach this site from the small town of Lennep located southeast of White Sulphur Springs on Montana Highway 294. From Lennep turn left onto the road to the ghost town Castle Town (Forest Road 581). Drive 10.4 miles on FR 581, passing Castle Town, to Hensley Creek and turn left on the small dirt road following the creek. Drive 0.7 mile to the gate and park. The mine tailings are beyond the gate and to the left. *DeLorme: Montana Atlas & Gazetteer:* Page 41 A8.

Rockhounding

Prospecting in the Castle Mining district southeast of White Sulphur Springs began in 1881, but it wasn't until several years later that mining took hold. A considerable amount of production took place with silver, lead, zinc, copper, and manganese being the contributing metals. The ore deposits are of several types, including fissure veins, replacement mineralization, and contact mineralization between Paleozoic limestone and igneous rocks. It has been determined by geologists that the mineralization took place as several episodes, each responsible for the deposition of certain primary elements.

Today, the small ghost town of Castle Town and numerous tailings piles are the only things left to tell of the mining days, and both are located off the beaten path. The most interesting material to collect is the malachite, which seems to be abundant at this particular site. The best place to look for malachite of cutting quality is at the dead end of the road and near the creek bed. The tailings piles are scattered around the end of the road and up the surrounding hills.

Lemhi Pass

See map on page 135
Land type: Mountains.
GPS: N. 44.97313 / W. 113.44321
Best season: Late summer.
Land manager: USDA Forest Service, Beaverhead-Deerlodge National Forest.
Material: Forsterite crystals, petrified wood.
Tools: Rock hammer.
Vehicle: Any.
Accommodations: Camping and RV parking in Beaverhead-Deerlodge National Forest; camping, RV parking, and motels in Dillon.
Special attractions: A monument to Sacagawea sits at the top of Lemhi Pass amongst striking high-mountain scenery. In 1805 Sacagawea led the Lewis and Clark expedition over the Lemhi Pass into Idaho.
Finding the site: From Grant drive west on Montana Highway 324 for 10 miles, turn right onto the road leading to Lemhi National Historic Landmark (Forest Road 300), and drive 12 miles to Lemhi Pass. The forsterite is located in the area around the pass. A good place to begin the search is on the road that leads from the pass to the Sacagawea monument. Drive a few hundred feet and search the hills on both sides of the road.

It should be noted that the road over Lemhi Pass is inaccessible most of the year due to snow. The road normally opens in July and remains accessible until it is snowed out. Inquire locally before trekking up to this site. *DeLorme: Montana Atlas & Gazetteer:* Page 20 A1.

Rockhounding

Forsterite is a mineral of the olivine group that can form gem-quality translucent green crystals. At this particular location, however, the forsterite is chiefly a yellowish color and forms in very large (usually a little smaller than 1 inch) specimens with little translucency. The crystals are abundant and can be found easily by searching the loose pieces of the country rock scattered on the hillsides. Although the museum-quality display may be out there waiting to be found, be prepared to be more fascinated with the large chunks of petrified wood that scatter the area than the forsterite. The wood doesn't appear to be in any particular location and its occurence is quite sporadic, but large colorful pieces can be found.

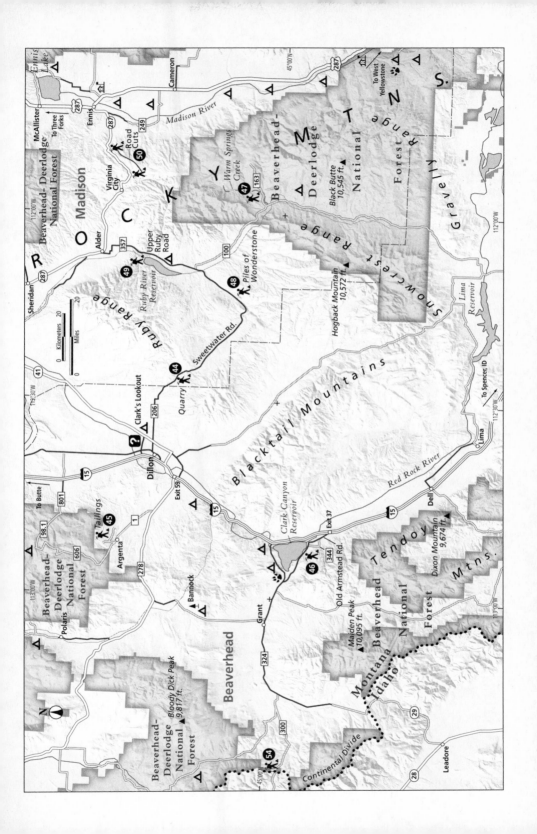

Spencer, Idaho

See map on page 137
Land type: High desert.
GPS: Large and sporadic area.
Best season: Generally open May through September.
Land manager: Private, Spencer Opal Mines.
Material: Opal.
Tools: None.
Vehicle: Any.
Accommodations: Camping, RV parking, and motels in Dillon, Montana, and in Idaho Falls, Idaho.
Special attractions: None.
Finding the site: From Interstate 15 at the Montana/Idaho border, drive south into Idaho for about 15 miles to the small town of Spencer. The mine is located just off the interstate on a small road (Old Highway 91). Follow the signs from the interstate. *DeLorme: Montana Atlas & Gazetteer:* Page 21 D8.

Rockhounding

The opal found in Spencer, Idaho, formed in small gas cavities and cracks within the volcanic rock. After the rhyolite and lava rock had already formed, a series of heated water passed through the rock, depositing siliceous material into the cavities.

The Spencer Opal Mine produces gem-quality opal that is often found in very thin layers. It's popularly used to make layered stones of thin opal and clear quartz. These "doublets" or "triplets" are extremely attractive and a nice display of them is located within the store, which is also a small rock shop with an extensive selection of Spencer opal jewelry. The store has a large pile of rock from the mine that is open to rockhounds to pick through for a very small fee. Trips to actually go into the mining premises to dig your own material are on selective weekends from May through September. They usually require a reservation well in advance.

For more information contact Spencer Opal Mines, HCR 62 Box 2060, Spencer, ID 83446; (208) 374–5476; www.spenceropalmines.com.

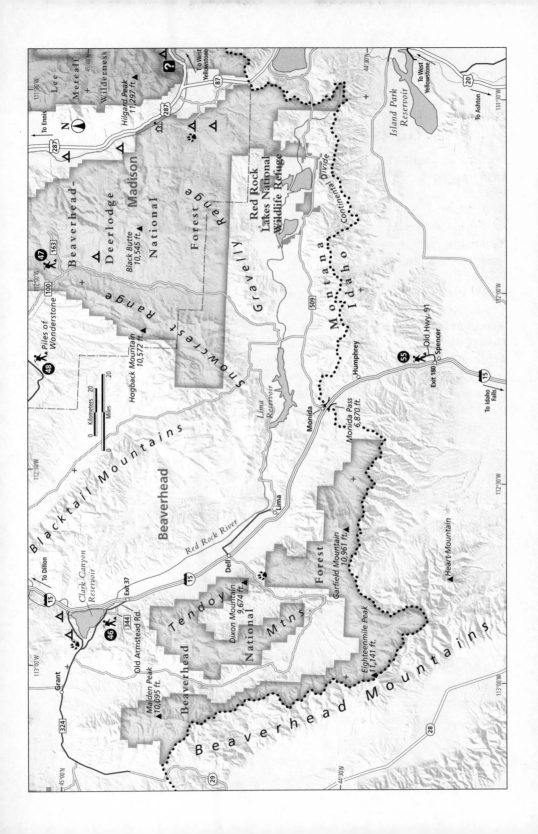

Gallatin Petrified Forest

See map on page 140
Land type: Mountains.
GPS: N. 45.22133 / W. 111.13211
Best season: Spring through fall.
Land manager: USDA Forest Service, Gallatin National Forest.
Material: Petrified wood.
Tools: Rock hammer.
Vehicle: Any.
Accommodations: Camping and RV parking on-site in Gallatin National Forest; camping, RV parking, and motels in Gardiner and Livingston.
Special attractions: Yellowstone National Park.
Finding the site: From Gardiner drive north on U.S. Highway 89 for 16 miles to just past the town of Miner. Turn left onto Tom Miner Basin Road, stay left at the T-intersection, and continue about 10 miles to the Tom Miner Campground. The trailhead is located at the bend in the campground, near the head of a beautiful mountain valley. Petrified wood can be found in the gullies and drainage of the area after hiking 1.5 miles, but the spectacular petrified tree stumps and "forests" are found several miles farther up the trail. A permit is required to collect petrified wood in the Gallatin Petrified Forest. The free permit can be obtained from the Forest Service office in Gardiner. Since it can take several weeks to receive a permit requested by mail, it would be wise to apply for one well in advance of any trip to the area. A permit can also be obtained in person by visiting the office during their hours of operation. It is highly suggested that you visit the office even if you already have your permit. The staff gives good directions, a map, and recommendations on hiking and collecting areas.

For more information send a request to Gardiner Ranger District, Box 5, Gardiner, MT 59030; (406) 848–7375. *DeLorme: Montana Atlas & Gazetteer:* Page 27 D6.

Rockhounding

The fossil forests of Yellowstone National Park are unique in that they lie one atop another. Entire forests were buried by volcanic ash and stream-carried volcanic sediment. New forests would grow, only to be buried by the same

Fossil tree stump at Tom Miner Basin.

processes. In time, the woody fibers of the trees in these buried forests were replaced by silica, thereby preserving the delicate cellular structures of the wood. A northern extension of these rock forests of Yellowstone can be found in southern Park and southeastern Gallatin Counties. This is the area known as the Gallatin Petrified Forest.

Wind, water, and ice have carved into the thick layers of volcanic rock in which the many petrified trees have been preserved. The logs and stumps are now exposed in the cliffs and steep slopes high on the valley sides. Petrified wood no longer can be found near the campground. It was removed long ago by eager collectors. Climbing to the higher elevations and carefully searching the rocky slopes north of the campground near Ramshorn Peak will reveal pieces that can be collected, as well as some very large petrified logs and stumps that can be photographed. It is necessary to hike a fair distance over rough and relatively high elevation terrain (more than 5,000 feet) to reach good collecting and viewing areas, so collectors should be in good physical condition.

Much of the fossil wood's state of preservation is extremely good. Frequently, the casts of limbs and logs prove to be hollow and lined with beautiful quartz crystals and, on occasion, crystals of amethyst. Some rockhounds

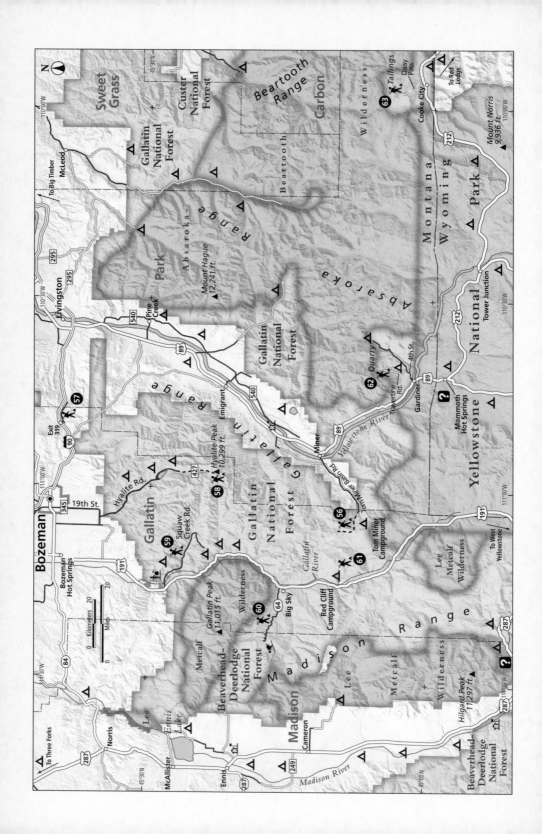

have reported the detailed impression of leaves in the finer grained siltstones sometimes associated with the coarser grained rock preserving the fossil wood.

Attempts have been made in the past by greedy collectors to remove large stumps and logs. When these attempts failed, the majestic monoliths were destroyed with hammer and chisel. It is hoped that the established collecting limitations will discourage this kind of ignorance.

The federal regulations concerning the amount of petrified wood that can be collected are different from those on other public land and are strictly enforced. Only a relatively small piece of wood, 20 cubic inches, can be collected by each individual per year within the Gallatin National Forest. Exercising extreme selectivity in your collecting is encouraged. The size of the "souvenir" specimen should in no way deter one from visiting the Petrified Forest, however. The number of standing stumps and fallen logs of petrified wood are worth seeing. A camera can record for you the fantastic part of Earth's geologic history that is preserved in the rocks at this unique area.

Bozeman Pass

See map on page 140
Land type: Mountains, road cut.
GPS: N. 45.66491 / W. 110.80243
Best season: Any.
Land manager: Gallatin County.
Material: Calcite crystals, heulandite, laumontite.
Tools: Rock hammer.
Vehicle: Any.
Accommodations: Camping, RV parking, and motels in Bozeman.
Special attractions: The Museum of the Rockies, on the grounds of Montana State University in Bozeman.
Finding the site: From Bozeman drive east on Interstate 90 for 12 miles and take exit 319 (Jackson Creek Road) to the frontage road on the south side of the interstate. Drive 1.6 miles east to the first road cut. *DeLorme: Montana Atlas & Gazetteer:* Page 27 A7.

Rockhounding

Zeolite minerals are popular with many collectors and refer to any of the minerals of the zeolite group. Several zeolite minerals as well as calcite can be found within exposures of the Livingston Formation in the vicinity of Bozeman Pass.

Zeolite minerals from the Bozeman Pass.

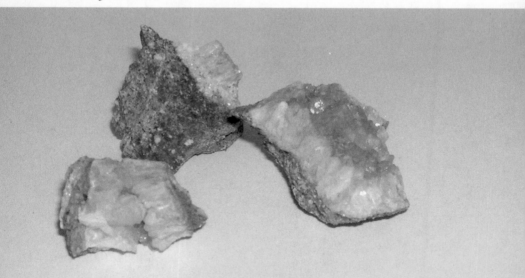

Small calcite crystals, heulandite, and laumontite crystals are exposed in the rocks here. Unfortunately most of them occur in the area along the interstate. It is unlawful to park along any interstate except in cases of emergency, so collecting is limited to the frontage roads near the pass. With diligent searching of the road and railroad cuts east of the pass, small zeolite mineral samples can be found. The minerals are not especially abundant, but by breaking some rock, a serious collector should be rewarded with specimens that may be suitable for micromount and possibly thumbnail collections. At the particular location mentioned, nice samples of heulandite can be found.

Hyalite Canyon

See map on page 140
Land type: Mountains.
GPS: N. 45.44768 / W. 110.96234
Best season: Late spring through fall.
Land manager: USDA Forest Service, Gallatin National Forest.
Material: Hyalite.
Tools: Rock hammer.
Vehicle: Any.
Accommodations: Camping and RV parking in Gallatin National Forest; camping, RV parking, and motels in Bozeman.
Special attractions: The Museum of the Rockies, on the grounds of Montana State University in Bozeman.

The beautiful scenery of Hyalite Canyon.

Finding the site: From Interstate 90 in Bozeman, take the 19th Street exit and drive south for 9.4 miles. Turn left onto Hyalite Road and drive 13.5 miles to the Hyalite Creek trailhead. Park and hike up Trail 427 for about 7 miles to the top of the area around the peak. *DeLorme: Montana Atlas & Gazetteer: Page 27 C7.*

Rockhounding

This site is for the motivated rockhound, and the trail requires a somewhat strenuous 7-mile hike in each direction. The scenic canyon derives its name from the rare presence of hyalite opal found in the rocks on Hyalite Peak at the head of the canyon. The opal is transparent to translucent and occurs on some of the rock exposed in the area. Unfortunately, the material seems to occur only in the area near the peak, and considerable searching and breaking of rock may be required to obtain specimens. Reaching the locality is a task, but for those ambitious enough to try, the scenery alone is worth the effort.

Squaw Creek

See map on page 140
Land type: Mountains, road cut.
GPS: N. 45.43648 / W. 111.17947
Best season: Spring through fall.
Land manager: USDA Forest Service, Gallatin National Forest.
Material: Quartz, jasper.
Tools: None.
Vehicle: Any.
Accommodations: Camping and RV parking in Gallatin National Forest; camping, RV parking, and motels in Bozeman and Big Sky.
Special attractions: The Museum of the Rockies, on the grounds of Montana State University in Bozeman.
Finding the site: From Bozeman drive south on U.S. Highway 191 for 16.5 miles. At the Squaw Creek Ranger Station, turn left onto Squaw Creek Road, cross the immediate bridge, and drive 4.5 miles east to the site. The andesite containing the drusy quartz is eroding from the left side of the mountain following the road. *DeLorme: Montana Atlas & Gazetteer:* Page 27 B5.

Rockhounding

Rubble andesite from Eocene time covers the mountains surrounding the Squaw Creek area and much of the southern Gallatin Range in general. The same andesite volcanoes that produced the rocks containing the drusy quartz were also responsible, at some point in time, for the complete mudflow burials of the area that produced Gallatin Petrified Forest.

The drusy quartz, referring to the pockets in the andesite completely covered in tiny quartz crystals, is abundant and covers the large volume of andesite rubble in the area. The rocks are often too large to move. The majority of the rocks are already split, and the quartz crystals, averaging less than ¼ inch in size, are exposed in their uninterrupted layers on the rocks. The crystals serve more as a thin sparkly blanket on the large rocks and wouldn't make suitable micromounts. This is an excellent site to collect exquisite garden rocks and samples for bookends or other rock-saw items, but quality above isn't available. While searching the hills keep an eye out for yellow and red jasper, which can also be found in the area.

Big Sky

See map on page 140
Land type: Mountains, road cut.
GPS: N. 45.27909 / W. 111.35172
Best season: Spring through fall.
Land manager: Montana DOT.
Material: Fossils.
Tools: Rock hammer.
Vehicle: Any.
Accommodations: Camping, RV parking, and motels in Big Sky.
Special attractions: Yellowstone National Park.
Finding the site: From the junction of U.S. Highway 191 and Montana Highway 64 in Big Sky, drive 5.9 miles west on MT 64 toward the Lone Mountain

Big Sky plant fossils.

Ski Resort Area to a large road cut and park. *DeLorme: Montana Atlas & Gazetteer:* Page 26 C4.

Rockhounding

The Cretaceous sandstone and shale in the road cut on the way to the Lone Mountain Ski Resort can yield nice specimens of plant fossils. The fossils are not very detailed and are preserved as black carbon impressions in the rock. They are not abundant, but loose specimens can be found at the base of the cut, and the shale breaks easily with a rock hammer.

Red Cliff

See map on page 140
Land type: Mountains, caves.
GPS: N. 45.17033 / W. 111.21881
Best season: Spring through fall.
Land manager: USDA Forest Service, Gallatin National Forest.
Material: Calcite crystals.
Tools: Rock hammer.
Vehicle: Any.
Accommodations: Camping and RV parking on-site; camping, RV parking, and motels in West Yellowstone and Big Sky.
Special attractions: Yellowstone National Park.
Finding the site: From Big Sky drive south on U.S. Highway 191 for 6.5 miles and turn left at the sign to Red Cliff Campground. Drive to the south end of the campground and park at the trailhead. Hike up the trail toward the rock cliffs and caves. The first collecting location is the most easily accessible one. It's located just a couple hundred feet up the trail in the first cave. Although only a short hike, it's extremely steep and should be attempted only by agile adult hikers. *DeLorme: Montana Atlas & Gazetteer:* Page 27 D5.

Rockhounding

Impressive specimens of golden calcite crystals can be collected at the Red Cliff Campground. Some of these translucent crystals are doubly terminated, and specimens of the larger clusters are often sold at a hefty price in local rock shops and Internet sites worldwide. This is a unique location since quality crystals that are several inches long can be collected if the amateur rockhound is able to break them free from the parent rock. The area has had quite a lot of collecting pressure over time, and the larger crystals won't be collected easily from the first cave. The rock containing the crystals occurs throughout the area around the steep trail, and further exploration of cavities in caves of the cliffs in the area is recommended. Either way, after visiting this locality you should be able to come home with at least a nice specimen to add to your collections. It should be noted that seasonally bats and other forest creatures call the Red Cliff caves home, and caution should be exercised to ensure their delicate habitat and quiet slumber, as well as your own safety.

Gardiner

See map on page 140
Land type: Mountains, road cut.
GPS: N. 45.04645 / W. 110.70839
Best season: Spring through fall.
Land manager: USDA Forest Service, Gallatin National Forest.
Material: Travertine.
Tools: Rock hammer, chisel, goggles.
Vehicle: Any.
Accommodations: Camping, RV parking, and motels in Gardiner.
Special attractions: Yellowstone National Park.
Finding the site: From Gardiner drive northeast on 4th Street for about 1 mile. At the fork in the road, turn left onto Travertine Road, drive 1.3 miles, and turn left and explore the road cuts. Do not trespass into the active mining areas; there is plenty of material to be found along the road. *DeLorme: Montana Atlas & Gazetteer:* Page 27 D8.

Rockhounding

Although Gardiner is most noted for being the north entrance to Yellowstone National Park, the vicinity has some interesting geological features. One of these is an extensive deposit of travertine, closely related to that being formed by hydrothermal activity at Mammoth Hot Springs south of Gardiner, just within the boundary of the park. Travertine is the name given to the layered, vuggy limestone that owes its origin to hot-spring deposition. It forms when the hot water carrying dissolved calcium carbonate reaches the surface, cools, and evaporates, thereby causing the mineral material to precipitate. Over long periods of time, thick deposits of travertine can accumulate.

Several sites are presently being quarried, and the material is shipped out to factories where it is shaped into attractive blocks and pieces for buildings, patios, etc. Access to active quarries is of course questionable, but road cuts high in the hills surrounding the quarries can provide rock-garden-variety pieces with less color. The collectible pieces tend to be white, red, orange, and pink banded, some with nice pockets or "vugs" of small calcite crystals. The travertine cuts nicely with a rock saw, and attractive bookends with vugs of crystals can be made.

Daisy Pass

See map on page 140

Land type: Mountains.

GPS: N. 45.05111 / W. 109.95163

Best season: Late spring through summer.

Land manager: USDA Forest Service, Gallatin National Forest.

Material: Pyrite, magnetite, chalcopyrite, garnet, quartz crystals.

Tools: Rock hammer.

Vehicle: High clearance, four-wheel drive.

Accommodations: Camping, RV parking, and motels in Cooke City.

Special attractions: Beartooth Pass, which crests at more than 10,000 feet, an elevation well above the tree line, and Grasshopper Glacier, where you can view locusts that were trapped in the glacial mass.

Finding the site: From Cooke City the New World Mining district can be reached via several rugged mountain roads. The 5 miles from Cooke City up to Daisy Pass are barely negotiable by regular automobile. A high-clearance four-wheel-drive vehicle is strongly recommended for the roads in this area. Some of the most scenic country of the nation is situated here. The trails leading from Cooke City to the old mining areas offer access to those who wish to explore the Beartooth Mountains further by backpack. The easiest directions to the site are to drive to the east end of Cooke City and turn left, heading north, onto the road connecting to Daisy Pass road. Go about 5 miles to the first mountainside covered in mine tailings.

Another highly recommended and popular option for those without a proper vehicle is to rent an ATV at Cooke City Motor Sports, located at the west end of Cooke City. For more information contact Cooke City Motorsports, P.O. Box 1120, Cooke City, MT 59020; (406) 838–2231. The business also provides free maps of the area. *DeLorme: Montana Atlas & Gazetteer:* Page 28 D4.

Rockhounding

High in the mountain passes north of Cooke City exists an area in which silver and lead were discovered in relatively large amounts during the early 1900s. In the vicinity of Daisy Pass, numerous mine dumps and tailings tell of a period when mining activity was much more pronounced.

Exposed here are Precambrian metamorphic rocks overlain by lower Pale-ozoic sedimentary rocks and Tertiary volcanic rocks. Tertiary intrusions in the form of laccoliths, sills, and dikes occur, and the mineralization in the area is thought to be related to these. Dumps and tailings support considerable amounts of pyrite, although most of it is considerably weathered. Magnetite is found along with some copper minerals. On a sunny day the rocks sparkle with the various metallic minerals and crystals streaming down the mountains.

Malta

See map on page 154
Land type: High hills.
GPS: N. 48.59771 / W. 107.73199
Best season: Spring through fall
Land manager: U.S. Highway, maintained by Montana DOT.
Material: Calcite crystals, fossils.
Tools: Rock hammer.
Vehicle: Any.
Accommodations: Camping, RV parking, and motels in Malta.
Special attractions: The Phillips County Museum located in Malta is the official repository for specimens collected by the Judith River Dinosaur Institute.

Flattened baculites and calcite crystals can be found by splitting shale concretions.

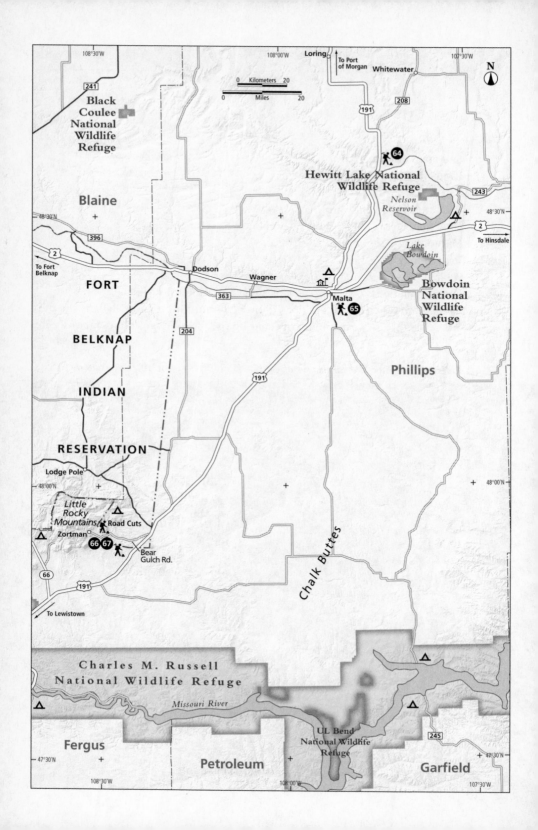

108°30'W · 108°00'W · 107°30'W

Loring
To Port
of Morgan
Whitewater

241
Black
Coulee
National
Wildlife
Refuge

191
208

64
Hewitt Lake National
Wildlife Refuge
Nelson
Reservoir
243

Blane

48°30'N
2
To Hinsdale

396
2
To Fort
Belknap

FORT

Dodson
Wagner
Malta
65

Lake
Bowdoin

Bowdoin
National
Wildlife
Refuge

363
BELKNAP

204
191

INDIAN

Phillips

RESERVATION

Lodge Pole
48°00'N

Little
Rocky
Mountains   Road Cuts
Zortman
66 67
Bear
Gulch Rd.

66
191

Chalk Buttes

48°00'N

To Lewistown

Charles M. Russell
National Wildlife Refuge
Missouri River

UL Bend
National Wildlife
Refuge

245

Fergus

Petroleum

Garfield

47°30'N

108°30'W · 108°00'W · 107°30'W

N

0   Kilometers   20
0   Miles   20

The museum has impressive paleontology exhibits along with displays on Montana history.

Finding the site: From Malta, drive north on U.S. Highway 191 for 19.5 miles to a large defined road cut and turn out where the material is located. *DeLorme: Montana Atlas & Gazetteer:* Page 91 B6.

Rockhounding

The Cretaceous Pierre Shale, a dark gray and brown shale, north of Malta, occurs in a very small section along the highway, but it's abundant in ammonites. The fossils, along with calcite crystals, are easily found by splitting the large shale concretions found on and around the hillsides.

Dinosaur Field Station

See map on page 154
Land type: Hills, badlands.
GPS: Large and sporadic area.
Best season: Programs offered during the summer, inquire for dates.
Land manager: Private, Judith River Dinosaur Institute (JDRI).
Material: Fossils.
Tools: All tools provided.
Vehicle: Any.
Accommodations: Camping, RV parking, and motels in Malta.
Special attractions: The Phillips County Museum located in Malta is the official repository for specimens collected by the Judith River Dinosaur Institute. The museum has impressive paleontology exhibits along with displays on Montana history.
Finding the site: The Dinosaur Field Station is located at the corner of U.S. Highway 2 and U.S. Highway 191 in Malta. The station is normally open to visitors seven days a week from May through September; inquire about off-season or special appointments. *DeLorme: Montana Atlas & Gazetteer:* Page 91 D5.

Rockhounding

For a real taste of the professional paleontology scene, Malta is the place to be. The Judith River Dinosaur Institute is dedicated to paleontological resource preservation, research, and public education. The Judith River Foundation is a nonprofit organization that operates the Dinosaur Field Station, which is open for tours during the summer months. For a small fee, the amateur rockhound can view a working professional paleontology lab.

Many significant paleontology finds have been discovered by JRDI and are often curated in the lab. One such famous fossilized character is the 77-million-year-old *Hadrosaur* (duck-billed dinosaur) "Leonardo." The dinosaur's biggest claim to fame is its unique preservation. Much of Leonardo's body is covered with fossilized soft tissue, including muscle, and rare skin impressions, and even the stomach contents remain. Often referred to as the "mummified" dinosaur, Leonardo was given the title of best-preserved dinosaur in the world by the *Guinness Book of World Records* in 2000.

On Dinosaur Field Station tours, the public is taken through a working

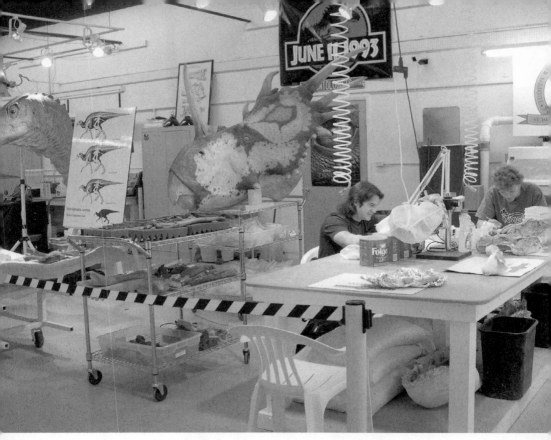

Volunteers prepare fossils inside the Dinosaur Field Station.

state-of-the-art paleontology lab. The staff describe geologic history of Montana, dinosaur excavation, and the workings of the lab. Yet, for the adventurous rockhound, a more hands-on experience is recommended. During the summer months JRDI invites the public to participate in dinosaur field excavations and exploration programs. The pay-digs are usually offered in five-day packages. The professional atmosphere of JRDI lends a rare chance for the novice to live the life of the scholarly dinosaur hunter. Typically, field-study packages include excavating, mapping, prospecting, and identifying dinosaurs. No experience is necessary, and the structured organized programs guarantee the best educational experience you can get digging up dinos. Reservations are needed, and participation is generally limited to children age fourteen and older. Each program is different, so inquire well in advance. For more information contact Nate Murphy, Judith River Dinosaur Institute, P.O. Box 429, Malta, MT 59538; (406) 654–2323; www.montanadinosaurdigs.com.

The Little Rockies

See map on page 154

Land type: Mountains, river bed.

GPS: See directions below.

Best season: Late spring through fall.

Land manager: Private.

Material: Gold, garnets.

Tools: Gold pan, trowel, flask.

Vehicle: High-clearance.

Accommodations: Camping and RV parking in Little Rocky Mountains; small motel in Zortman; camping, RV parking, and motels in Malta.

Special attractions: Fossil Hill, which is full of belemnites, oysters, small ammonites, and calcite crystals. Locals will help you acquire access to this private land.

Finding the site: Just northeast of the junction of Montana Highway 66 with U.S. Highway 191 (between Malta and Lewistown), turn west onto the road leading to Zortman. Drive 7 miles, staying left and following the signs to Zortman. The road that leads you to town will take you straight to the Zortman Garage and Motel, which also hosts a small rock shop that sells gold pans, tools, and various supplies. Inquire within the motel office about land status and locations. *DeLorme: Montana Atlas & Gazetteer:* Page 74 B2.

Rockhounding

Placer gold was first discovered in this area in 1884, but ten years of continuous placer operations took place before the lode deposits containing the gold were mined. Apparently, the alluvial deposits in which the gold was first discovered are very extensive, but the quantity of water needed to work these deposits is not available. Lode deposits have since provided the major amount of recoverable gold in the district. Several mines near Zortman and Landusky have been reactivated and there are many claims in the area.

The unusual occurrence of the Little Rockies, a small rugged mountain mass in the plains of northern Montana, is attributed to volcanism. The Little Rockies are basically interpreted as a laccolithic structure, that is, an area domed upward as a result of molten magma intruding between sedimentary rock layers and pushing them from below. A later period of erosion has exposed the igneous rock core and the metalliferous veins it contains, with gold and silver being the main metals of interest.

Fossil Hill

See map on page 154
Land type: Low mountains, road cut.
GPS: N. 47.91119 / W. 108.39698
Best season: Any.
Land manager: Phillips County Road.
Material: Fossils.
Tools: None.
Vehicle: Any.
Accommodations: Camping and RV parking in Little Rocky Mountain, small motel in Zortman; camping, RV parking, and motels in Malta
Special attractions: Gold panning in Zortman. The Zortman Garage and Motel has a rock shop that sells equipment and supplies. Also inquire there on the status of Fossil Hill, an area behind the Zortman dump that is chock-full of belemnites, oysters, small ammonites, and calcite crystals. This site is popular with universities, who have conducted digs that have uncovered important dinosaur fossils on several occasions.
Finding the site: The first road cut is on the way to Zortman. Just northeast of the junction of Montana Highway 66 with U.S. Highway 191 (between Malta and Lewistown), turn west on to the road leading to Zortman. Drive about 5 miles, park, and explore the limestone road cuts for oysters, belemnites, and other small marine fossils. *DeLorme: Montana Atlas & Gazetteer:* Page 74 B2.

Rockhounding

This location can satisfy the paleontology rockhound as well as someone who has the "gold fever." The best collecting sites are on private property, and locals are helpful in acquiring access to Fossil Hill. The key to collecting in these locations is familiarization with the Madison limestone. It's not hard to locate the light-colored rocks that surround the mountain range. Paleozoic and Mesozoic rock layers are turned up sharply along the margins of the Little Rockies, and exposures of various fossiliferous formations often occur in the area. The marine Jurassic shales here continue to yield relatively common specimens of the oyster *Gryphaea,* the belemnite *Pachyteuthis,* the star-shaped crinoids *Pentacrinus,* as well as less common species of ammonites and brachiopods.

Bear Creek

See map on page 162
Land type: Hills.
GPS: N. 47.94766 / W. 106.29301
Best season: Spring through fall.
Land manager: Bureau of Land Management (BLM).
Material: Fossils, calcite crystals.
Tools: Rock hammer.
Vehicle: High clearance, four-wheel drive when wet.
Accommodations: Camping and RV parking on Fort Peck; Camping, RV parking, and motels in Glasgow.
Special attractions: Fort Peck Dam Interpretive Center and Museum has displays on Fort Peck Dam as well as geology, with an impressive paleontology exhibit and life-size T-Rex models. Fort Peck Dam is one of the largest earth-filled dams in the world.
Finding the site: From Fort Peck drive south on Montana Highway 24, cross the dam, drive 13.5 miles, and turn right onto Bear Creek Road. If the gate is closed, there may be cattle loose so be sure to close it behind you after driving in—the public is allowed to enter the area. Drive about 2 miles down the road to the Bear Creek entrance to the shore of Fort Peck Lake. Park before you enter the Recreational Area or cross into the fence separating the BLM land from the Recreation Area, and stay on the side of the fence away from the lake while rockhounding. A BLM map is recommended. *DeLorme: Montana Atlas & Gazetteer:* Page 77 B5.

Rockhounding

Several barriers keep the rockhound from collecting on Fort Peck Lake, home to some of the best marine fossil specimens to be found in the state. The main reason is that vertebrate collecting of any kind is illegal by federal law, and the law is firmly enforced within the Fort Peck area. Collecting rocks of any kind is prohibited within the Charles M. Russell National Wildlife Refuge, and there are signs posted everywhere warning that violators will be prosecuted and can face jail time for disobeying. One might think this is to preserve the refuge in its natural state, but a rockhound may look at it as a sign of how spectacular the fossils are that wash up on shore. The shoreline of this area,

A life-size model of T-rex greets visitors inside the Fort Peck Dam Interpretive Center and Museum.

particularly at Bear Creek, is intermingled with eroded Cretaceous Bearpaw Shale, and all the mollusks can be found along the shoreline loose in the silt or in the rocks along the lakeshore.

The solution to the collection limitations is to find the patch of BLM land closest to the lakeshore. The Bear Creek BLM land that borders the Fort Peck Bear Creek Bay Recreation Area offers good opportunity, but diligent searching is required. The best place to begin your search is by parking your car at the fence line on the BLM side and walking to the hills toward the east. At 0.25 mile into the field, look for shale concretions that are eroding from the hills. Finding fossils doesn't come easily, and the site requires prospecting for large enough pieces of the dark gray Bearpaw Shale to split. Nevertheless, whole ammonites and mollusks can be found, although they are often flattened. Also be on the lookout for calcite crystals that can be found in pockets in the concretions.

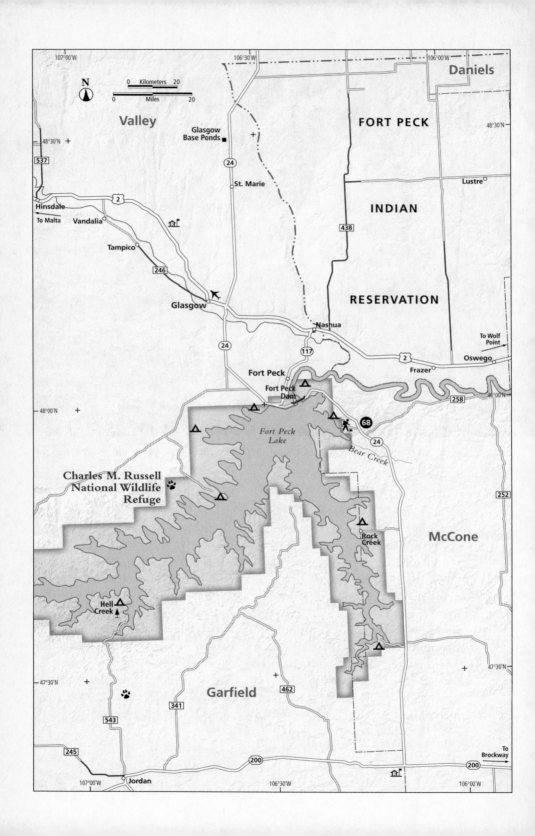

The area around Fort Peck Lake provides some of the best exposures of Cretaceous Bearpaw Shale, Judith River, and Hell Creek Formations to be found in the state. In the early 1900s Barnum Brown, a paleontologist for the American Museum of Natural History, discovered in the 65-million-year-old Cretaceous sediments of the Hell Creek Formation the skeletal remains of a remarkable creature—the *Tyrannosaurus rex,* which has come to epitomize the dinosaur. Skeletons of the armored, three-horned dinosaur *Triceratops* are also frequently found here, as well as the remains of *Hadrosaurs* and the rare *Ankylosaurs.*

The sediments of the Hell Creek Formation southeast of Fort Peck Lake have produced more than just dinosaurs. Numerous skeletal remains of tiny mammals have been collected from several badlands areas as well.

Yellow Water Reservoir

See map on page 165

Land type: Low hills, lakeshore.

GPS: N. 46.90007 / W. 108.47385

Best season: Late spring through fall.

Land manager: Montana Fish, Wildlife, and Parks.

Material: Fossils.

Tools: Rock hammer.

Vehicle: Any.

Accommodations: Camping, RV parking, and motels in Roundup.

Special attractions: None.

Finding the site: From the junction of Montana Highways 200/244 in Winnett, drive 7.5 miles south on MT 244 and turn right onto a gravel road leading to the Yellow Water Reservoir. Drive 6.1 miles to the lake and the various limestone bluffs that meet the shore. A good collecting area is the rocky outcrop on the south edge of the reservoir near the lakeshore, just after you cross the dam. *DeLorme: Montana Atlas & Gazetteer:* Page 60 C2.

Rockhounding

This site is a lot of fun, but navigating it requires a lot of map reading. More than half of the reservoir is located on part of the War Horse National Wildlife Refuge; a simple map of the area is required to ensure that you do not collect any rocks on the protected refuge land.

A conspicuous sandstone ledge and cuesta parallels the southern shoreline of the reservoir for some distance. The ledge is formed in the Mosby Sandstone member of the Cretaceous Colorado Shale, and in some spots it's very fossiliferous. Sand concretions can be found that contain abundant snails, some oysters, and occasionally a rare specimen of the relatively large ammonites. Be prepared to use the rock hammer, as the best specimens may require splitting layers and concretions, although there are some loose samples.

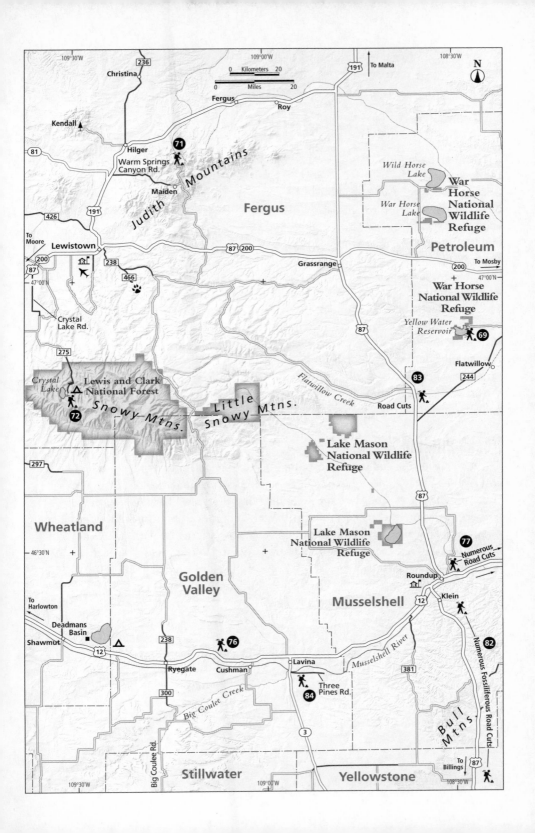

Sidney

See map on page 167
Land type: River, gravel bars.
GPS: N. 47.67382 / W. 104.15707
Best season: Spring through fall.
Land manager: Montana Fish, Wildlife, and Parks.
Material: Agate, petrified wood, jasper.
Tools: None.
Vehicle: Any.
Accommodations: Camping, RV parking, and motels in Sidney.
Special attractions: The Mondak Heritage Center in Sidney has displays on Montana history and pioneer life, including an indoor exhibit of a reconstructed pioneer town.
Finding the site: From Montana Highway 200 about 2.5 miles south of Sidney, turn left onto Montana Highway 23. Drive 1.5 miles, cross the Yellowstone River, and turn left into the fishing access. Park and explore the gravel bars. *DeLorme: Montana Atlas & Gazetteer:* Page 79 C8.

Rockhounding

The best agate collecting in the Sidney fishing area seems to be upstream, where gravel bars are more exposed. This area has an abundance of material if you are willing to spend the time walking along the gravel bars. As you move farther upstream, the pieces become larger and more abundant. Most are translucent and many contain beautiful moss agate patterns. As with many sites near water in eastern Montana, be prepared for mosquitoes the size of bombers.

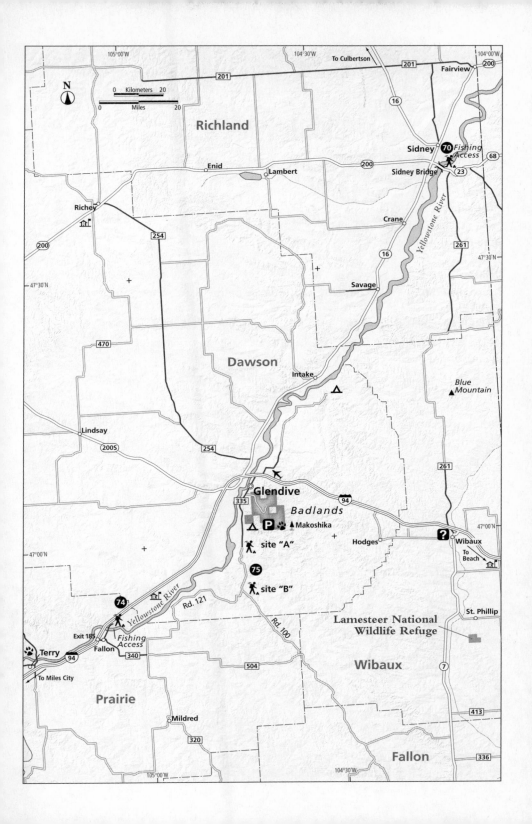

Judith Peak

See map on page 165
Land type: Mountains.
GPS: N. 47.21768 / W. 109.22389
Best season: Spring through fall.
Land manager: Bureau of Land Management (BLM), private.
Material: Quartz crystals.
Tools: None.
Vehicle: Any.
Accommodations: Open camping on BLM; camping, RV parking, and motels in Lewistown.
Special attractions: The Central Montana Museum in Lewistown has displays on Montana history and a small collection of fossils and minerals.

The spectacular valley view from the radar tower where the "Montana Diamonds" are located.

The doubly terminated quartz crystals have a naturally occurring diamond shape and are located in the light porphyritic rock in the road cuts.

Finding the site: From Lewistown drive north on U.S. Highway 191 for 10 miles and turn right onto Warm Springs Canyon Road, following the signs toward the ghost town of Maiden. Drive 9.4 miles and turn left, following the signs to Judith Peak. Continue 3.1 miles to the first good outcrop of a light feldspar rock. You can begin searching at this point. Another good collecting area is located down the faint road on the right; follow it to the dead end, which is a couple of hundred feet after the turnoff. At this location an abundance of crystals were found in the soil at the base of the outcrop. If you continue about 1 mile up the road to Judith Peak, you will reach a radar tower. This area offers a stunning view and the same outcropping with Montana Diamonds, along with large veins of smoky quartz and the occasional crystal pocket. The summit of the peak was at one time the site of a U.S. Air Force radar station. The view from the old radar tower is well worth the short drive. On a clear day most of the nearby mountain systems can be viewed, as well as several interesting geologic features on the prairie to the east. *DeLorme: Montana Atlas & Gazetteer:* Page 59 A7.

Rockhounding

Judith Peak is located at 5,808 feet and towers above the valley below. This site is both an amazing place to visit with a spectacular valley view and an excellent site for collecting one of Montana's unique rockhounding treasures.

Small doubly terminated crystals of quartz often referred to as "Montana Diamonds" weather from the porphyritic rock containing them; these are sought by mineral collectors primarily because of their extremely uniform shape. Although interesting and very abundant, they lack the quality of the famed Herkimer "diamonds" of New York and New Jersey. The crystals are not clear but are generally a very dark smoky and opaque variety of quartz. Sizes range from about ¼ to 1 inch in diameter. The crystals have been used as jewelry items by some rockhounds because of their perfect symmetrical form. The porphyry is highly weathered, and the doubly terminated crystals are lying free in the soil and weathered rock in the large road cuts near the summit of the peak.

The dark quartz crystals are easily seen in the solid rock that composes the road cut, but they normally cannot be removed from the parent rock without being broken. The texture of the rock containing the crystals demonstrates its plutonic or intrusive nature. The best crystals are those that have weathered naturally from the rock and can be found loose in the gravels, but be selective because most are not of high quality. On any given day it is easy to go home with several dozen crystals found loose in the gravel. The sun offers great assistance to finding the quartz in the soil, as they are more easily seen glistening in the sunlight.

Crystal Lake

See map on page 165
Land type: Mountains, road cut, lakeshore.
GPS: N. 46.80424 / W. 109.51165
Best season: Summer through fall.
Land manager: USDA Forest Service, Lewis and Clark National Forest.
Material: Fossils.
Tools: Rock hammer.
Vehicle: Any.
Accommodations: Camping in the Big Snowy Mountains; camping, RV parking, and motels in Lewistown.
Special attractions: The Central Montana Museum in Lewistown has displays on Montana history and a small collection of fossils and minerals.
Finding the site: From Lewistown drive 8.5 miles west on Montana Highway 200 and turn left onto Crystal Lake Road. Follow the signs to Crystal Lake; it is about a 20-mile drive on well-maintained gravel roads. This road has several road cuts with promising leads; don't limit exploring to just around the lake. *DeLorme: Montana Atlas & Gazetteer:* Page 59 C6.

Rockhounding

The area in and around the Big Snowy Mountains contains fossils in the many outcrops of varying ages from the Early Paleozoic to Mesozoic. The Mississippian Madison limestone is well exposed, occurs throughout the range, and is said to have a strong presence of various marine fossils such as brachiopods, corals, and crinoids, where it is found. Relatively complete crinoids have also been found in some of the thin-bedded Madison limestone. They are difficult to find and are obviously highly prized.

The foothills of the range consist of upturned layers of Mesozoic rocks and marine fossils, such as belemnites, oysters, and crinoids. The Jurassic shales appear to be the best fossil-producing rock in the Mesozoic sequence, but good exposures are not common. There are many roads throughout the range that offer an excellent chance to explore this beautiful area a little longer.

A good place to start collecting is on the road leading to Crystal Lake, a popular camping area in the northwest part of the Big Snowy Mountains. The scenic lake is reason enough to choose this route. The road passes through

Nice specimens of marine fossils are found in the limestone gravels and large boulders around the area of stunning Crystal Lake.

upturned units of Paleozoic rocks and provides easy access to varying ages of rocks. A search of the road cuts and cliffs will reveal the fossiliferous nature of many of these beds. More so, however, the limestone gravels of the shore of Crystal Lake produce nice samples of small marine fossils. As you walk farther from shore, larger limestone boulders occur with more variety of fossils. This material is extremely hard to break, and the most efficient collecting is done by picking up pieces that have already been split from the larger rocks.

Yogo Creek

See map on page 174

Land type: Mountains, stream.

GPS: N. 46.89883 / W. 110.38785

Best season: Summer through fall.

Land manager: USDA Forest Service, Lewis and Clark National Forest.

Material: Fossils, gold.

Tools: Rock hammer, gold pan, trowel, fleck flask.

Vehicle: High clearance.

Accommodations: Camping and RV parking on-site in Lewis and Clark National Forest; camping, RV parking, and motels in Lewistown and Harlowton.

Special attractions: The Central Montana Museum in Lewistown has displays on Montana history and a small collection of fossils and minerals.

Finding the site: These directions are to one particular site, but there are several good panning sites and road cuts of fossiliferous limestone in the area.

From Sapphire Village, a small town off Montana Highway 239 southwest of Lewistown, take the road heading west out of Sapphire Village toward the Judith River Wildlife Management Area within the Lewis and Clark National Forest. Drive 3 miles and turn right onto Forest Road 266. The massive limestone outcrops are 5.5 miles down FR 266. To continue to the gold-panning area, stay on FR 266 and go right at the fork. Drive about 2 miles to the campsites on the left side of the road along Yogo Creek. Park at any site and explore the area for a good place to dig and pan along the creek. Some samples of loose limestone with fossils can be found along the creek as well. *DeLorme: Montana Atlas & Gazetteer: Page 58 C2.*

Rockhounding

The first thing you will probably wonder is why this isn't a sapphire site. Of all the famed gems in Montana, the Yogo sapphire from the Little Belt Mountains is the most precious. The deep cornflower blue color of the sapphires found here rivals even that of Ceylon verities. The sapphires of Yogo Creek are the only ones known in the world that naturally occur in a deep uninterrupted blue.

Yogo Gulch was originally a source of placer gold, and it was strictly by chance that the Yogo sapphire was discovered. Because of their relatively high

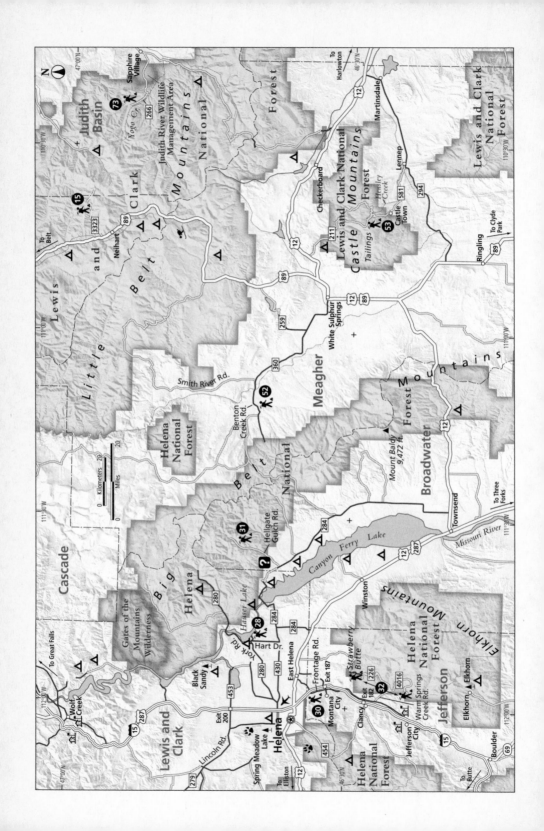

density, the unusual blue stones were common in the gold-bearing stream gravels. Curiosity resulted in an evaluation of the stones, and then placer operations were begun to retrieve sapphires along with gold. These operations continued up to the investigation and eventual mining of a large 4-mile-long igneous dike containing the sapphires. Today this sapphire mine is the only one in the world that mines the precious stones out of hard rock. Inaccessible to collectors, the mine is currently being developed, and there is a great deal of confidence that gems will be produced from this area for some time to come.

Fortunately, there are still precious treasures for the rockhound to collect in the public land along Yogo Creek. Decent amounts of gold can be found with diligent panning. If the minerals don't "pan out," there is a great deal of fossiliferous Mississippian limestone exposed throughout the Little Belt Mountains. The approximately 300-million-year-old rocks contain marine fossils that resulted from an ancient shallow sea that once covered the area. The samples to be found in the road cuts around Yogo Creek are small spirifer fossils in good condition. Unfortunately, and quite typically of limestone, the rock is difficult to break out of the cut. Your best luck will be had by searching already loose rocks on the ground. If you have a forest map and a good sense of adventure, it is documented that a highly fossiliferous reef of limestone is located on the south side of Bandbox Mountain just a few miles north of this site.

Fallon

See map on page 167

Land type: River, gravel bars.

GPS: N. 46.85632 / W. 105.11490

Best season: Spring through fall.

Land manager: Montana Fish, Wildlife, and Parks.

Material: Agate, jasper, chalcedony, petrified wood.

Tools: None.

Vehicle: Any.

Accommodations: Camping, RV parking, and motels in Glendive.

Special attractions: Makoshika State Park, a prime area to observe badlands topography.

Finding the site: From Interstate 94 near Fallon, take exit 185 to the frontage road on the south side of the interstate. Drive east on the frontage road for 1.5 miles, cross the bridge, and turn right into the boat launch and parking area. Walk to the river and search the gravel bars. *DeLorme: Montana Atlas & Gazetteer:* Page 64 C3.

Rockhounding

Like any agate-hunting locality in Montana, the absolute best time of year to collect is in early spring and late summer when the river is at its lowest point. The best locality at the Fallon boat launch area to collect agate is about a 0.25-mile walk north along the river to some large gravel bars.

A decent amount of Montana moss agate, red jasper, and smoothed petrified wood can be found here by walking slowly, preferably on a sunny day, along the gravel bars. There are reports of the popular blue-gray Montana agate being found here as well. On the particular time of my visit, a rare Montana Fairburn agate, an agate that has characteristic potato-eye swirls of color, was found. Besides the agate though, there is not too much going on in Fallon.

Glendive

See map on page 167
Land type: Badlands.
GPS: N. 46.90613 / W.104.74316
Best season: Spring through fall.
Land manager: Bureau of Land Management (BLM).
Material: Fossils, petrified wood, sandstone concretions, selenite crystals, barite crystals.
Tools: Rock hammer.
Vehicle: Any.
Accommodations: Camping and RV parking in Makoshika State Park; camping, RV parking, and motels in Glendive.

Doug Smith of Missouri River Travel searches for fossils in the eastern Montana badlands.

Ammonites from site A.

Special attractions: Sculptured by running water, Makoshika State Park near Glendive is a spectacular example of badlands topography. The badlands are composed mainly of soft shales and sandstones, and the sediments of the Tertiary Fort Union Formation and the Cretaceous Hell Creek Formation in eastern Montana are easily affected by the agents of weathering and erosion. The remains of fossil plants, dinosaurs, and small mammals, some exceedingly rare, have been removed from these beds. The Makoshika Dinosaur Museum in central Glendive hosts many paleontology displays.

Finding the site: Collecting anywhere within Makoshika State Park is prohibited and strictly enforced.

Some of the more interesting fossils worth searching for outside Makoshika State Park are the leaf prints in the sandstone and shale layers of the Fort Union Formation, and the fossil "figs" and reversed casts of pine cones in the Hell Creek Formation. The sandstone casts called "figs" have the same general shape as present-day figs, but their actual identity continues to be debated by paleo-botanists. If you stop by the visitor center at the entrance of the park, a map can be requested for the BLM land with the Hell Creek Formation where col-lecting is legal.

The first site (site A) is on BLM land and can be reached from Glendive by driving south on Main Street (Road 100/Montana Highway 335) for 8.5 miles until the road turns to gravel. At this point there should be a large parking area on both sides of the road predominantly for the Glendive Short Pine OHV Area. Park here and explore hillsides for marine fossils in the shale.

The next site (site B) is reached by driving 8 miles down Road 100 from the previous site. Veer left at the second fork (staying on Road 100) and the BLM badlands are on the south side of the road for several miles. Search this area for petrified wood, sandstone concretions, fossilized plants, sequoia cones, and leaf casts. *DeLorme: Montana Atlas & Gazetteer:* Page 65 B5 and C5.

Rockhounding

Site A: At the first site, outcrops of the Late Cretaceous Pierre Shale occur in road cuts, bluffs, and stream banks along Cedar Creek south of Glendive. Only the uppermost portion of this entire sequence of strata in the Pierre Shale is exposed, so the fauna that can be collected is limited in variety. Among the fos-sils to be found here are numerous pelecypods; several species of ammonites; many different genera and species of small gastropods; occasional nautiloids; and rare echinoids. Some of the limestone concretions near the top of the dark exposures of Pierre Shale consist almost entirely of pelecypods. Careful break-ing of the concretions may uncover excellent specimens of mollusks, many with their original mother-of-pearl shell still in tact.

Periodically, septarian concretions are found that contain unusual and attractive deposits of calcite and barite crystals. Lying free on the mounds of gray bentonite that occur throughout the area are abundant "fishtail twin" crystals of selenite and less common gray nodules composed of radiating barite crystals.

Concretions that have washed down from higher levels litter the creek bot-toms, and good fossils often can be found here. Large pieces of agatized wood frequent the gravels of the creek bottoms, but their source is uncertain.

Site B: Exposures of the Cretaceous Fox Hills and Hell Creek Formation directly to the east and west of the Pierre Shale outcrops along Cedar Creek are recognized by the presence of light-colored sandstones and greenish gray shales. These rock units contain fossils similar to those found in the Makoshika State Park area described earlier. Even though vertebrate fossils ranging from crocodile teeth to whole dinosaurs are found in this area, remember that vertebrate collecting is illegal on public land.

The rock units also contain some unusual reddish sandstone concretions, most likely with high iron contents, which occur as single spheres or groups of spheres. These make interesting additions to a rock collection and are found in gullies and ravines where they have weathered from the rock. Some exposures of the Fox Hills Sandstone just above the outcrops of Pierre Shale will provide specimens of a very unusual fossil. They are rust colored, cylindrical in shape, about 1 inch across, and vary in length up to about 1 foot. They are commonly referred to as "petrified corncobs," probably in reference to their bumpy exterior and general appearance, but they are actually the fossil casts of shrimp burrows.

Ryegate

See map on page 182
Land type: Hills, road cut.
GPS: N. 46.31284 / W. 109.12935
Best season: Any, when dry.
Land manager: Montana DOT.
Material: Fossils.
Tools: None.
Vehicle: Any.
Accommodations: Camping, RV parking, and motels in Roundup.
Special attractions: None.
Finding the site: From the intersection of U.S. Highway 12 and Montana Highway 300 in Ryegate, drive east on US 12 for 6.5 miles to a road cut on

Both halves of a fossilized oyster from Ryegate.

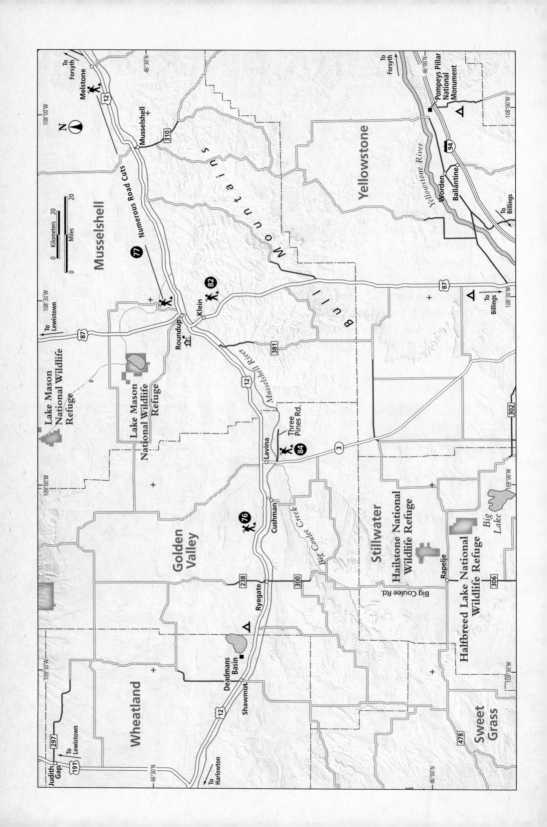

the north side of the road. Search the road cut for loose oysters or conglomerate rocks. The heaviest concentration observed was at the east end of the road cut. *DeLorme: Montana Atlas & Gazetteer:* Page 43 B8.

Rockhounding

Similar to the fossils found at Lavina, Cretaceous oysters at this particular site can be found easily in the road cut, although the fossils here are not as concentrated and the lignite layer is not found. The samples are easy to collect and most often complete. A rock hammer is not necessary because the fossils are loose.

Roundup to Melstone

See map on page 182
Land type: Low mountains, road cut.
GPS: Large and sporadic area.
Best season: Any.
Land manager: U.S. Highway, maintained by Montana DOT.
Material: Fossils.
Tools: Rock hammer, chisel.
Vehicle: Any.
Accommodations: Camping, RV parking, and motels in Roundup.
Special attractions: Musselshell Valley Historical Museum in Roundup has a small selection of fossils as well as historical relics.
Finding the site: This site consist of several miles of fossiliferous road cuts between Roundup and Melstone. From Roundup drive east on U.S. Highway 12 following road cuts of sandstone and shale to Melstone; the best location seems to be directly between the towns. *DeLorme: Montana Atlas & Gazetteer:* Page 44 A2–4.

Rockhounding

Searching through these road cuts is just as enjoyable as the scenic eastern Montana drive along this highway parallel to the Musselshell River. From Roundup you head east into the oil-pumping fields of Melstone and Ingomar, with of course, the veins of coal leading the way. Much like the site south of Roundup, carbon prints of trees, branches, leaves, and other organics can be found easily by splitting the sandstone and shale in the road cuts. Bring suitable packaging materials, as the prints can easily rub off.

Miles City

See map on page 186
Land type: Badlands.
GPS: N. 46.42359 / W. 105.67654
Best season: Late spring through summer.
Land manager: Bureau of Land Management (BLM).
Material: Petrified wood, iron concretions.
Tools: Rock hammer.
Vehicle: High clearance.
Accommodations: Camping, RV parking, and motels in Miles City.
Special attractions: The Range Riders Museum and Bert Clark Gun Collection in Miles City features a collection of western antiques, artifacts, and small geology displays.
Finding the site: From the junction of Interstate 94 and U.S. Highway 12, take US 12 east toward Baker for 5.6 miles. Turn left onto an unmarked dirt road near a large gravel lot and drive 0.7 mile into an area of badlands. Park and explore. *DeLorme: Montana Atlas & Gazetteer:* Page 47 A8.

Rockhounding

Around Miles City in the Fort Union Formation, logs, stumps, and limbs of fossilized sequoia trees periodically are found weathering in the hillsides. The gravels of some of the stream bottoms in the region contain relatively large fragments of colorful agatized petrified wood, ideal for rock gardens. Keep the government regulation concerning petrified wood in mind if you are on public land and decide to collect any of this material. In some cases, one or two pieces could easily fill the yearly quota.

Sandstone formations high in the badlands can produce wonderful specimens of petrified wood and iron concretions. Search in and around wash beds for petrified wood, which is generally rusty colored and quite large. This is a large site with much to explore, and a BLM map is recommended.

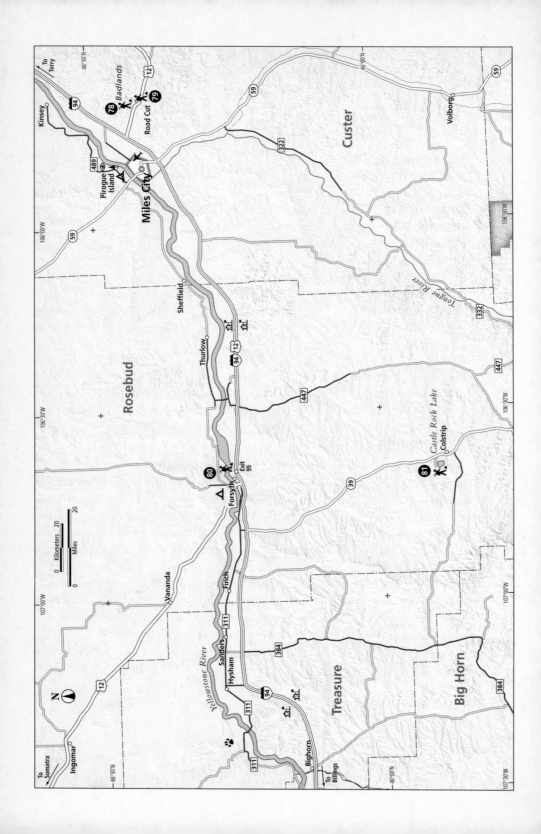

Fort Union Formation

See map on page 186
Land type: Hills, road cut.
GPS: N. 46.40567 / W. 105.66079
Best season: Any.
Land manager: U.S Highway, maintained by Montana DOT.
Material: Fossils.
Tools: Rock hammer, chisel.
Vehicle: Any.
Accommodations: Camping, RV parking, and motels in Miles City.
Special attractions: The Range Riders Museum and Bert Clark Gun Collection in Miles City features a collection of western antiques, artifacts, and small geology displays.
Finding the site: From the junction of Interstate 94 and U.S. Highway 12, drive east on US 12 for 6.5 miles to a large road cut of red clinker. *DeLorme: Montana Atlas & Gazetteer:* Page 47 A8.

Rockhounding

The Tertiary Fort Union Formation, which is so predominant in the eastern part of Montana, has been a prolific source of extremely well-preserved fossil leaf imprints. The imprints of hardwood trees such as birch, elm, oak, and maple are most common. Most exposures of the Fort Union Formation in highway road cuts have the potential of producing detailed fossils if the thin-bedded sandstones and hard shales are carefully split with a hammer and chisel.

The red clinker beds will disclose fine delicate impressions of leaves. These red beds represent clays and mudstones that were baked by the natural burning of coal layers underground. The highly baked clays that now outcrop along the crests of the hills in the area have the appearance of volcanic rock and are often referred to as "scoria." At this particular site there is plenty of material to be found by searching the piles of loose clinker along the base of the road cuts and splitting the larger pieces.

Forsyth

See map on page 186
Land type: Riverbank.
GPS: N. 46.27444 / W. 106.67907
Best season: Spring through fall.
Land manager: Montana Fish, Wildlife, and Parks.
Material: Agate, jasper, petrified wood.
Tools: None.
Vehicle: Any.
Accommodations: Camping, RV parking, and motels in Forsyth.
Special attractions: None.
Finding the site: From Interstate 94 take exit 95 toward Forsyth. Following the signs to the Rosebud Recreation Area, drive north and turn right onto 15th Street. Go 0.4 mile to the recreation area for the Yellowstone River. *DeLorme: Montana Atlas & Gazetteer:* Page 46 B3.

Rockhounding

The best and most efficient times to search for agate are in spring, just as the snow melts but the rivers are not yet full, and during late summer, as the river reaches the driest point. In Forsyth, search the banks, gravel bars, and shoreline for agate, jasper, and petrified wood. A sunny afternoon is also a good time to look for the transparent agate. As usual for the Yellowstone in late summer, the only things more abundant and seemingly larger than the agate are the imposing mosquitoes that seem to thrive off the potent repellent DEET.

Castle Rock Lake

See map on page 186
Land type: Hills, lakeshore.
GPS: N. 45.89809 / W. 106.64055
Best season: Spring through fall
Land manager: Colstrip Parks and Recreation District.
Material: Fossils.
Tools: Rock hammer, chisel.
Vehicle: Any.
Accommodations: Camping, RV parking, and motels in Forsyth.
Special attractions: None.
Finding the site: From Montana Highway 39 in Colstrip 1 mile north of the high school, turn left onto the dirt road leading west to Castle Rock Lake. Drive 0.4 mile to the Castle Rock Lake Recreation Area. From the parking area, walk the trail around the lake and search the cuts along the path for fossil imprints of various organics in the red clinker. *DeLorme: Montana Atlas & Gazetteer:* Page 46 D3.

Rockhounding

The area around Colstrip is famous for its large deposits of low sulphur coal. Deposits occur as relatively thick seams in the Tertiary Fort Union Formation and have been strip-mined for many years.

Exposures of the Fort Union Formation in the area contain well-preserved fossil leaf prints, similar to those found near Miles City. Although not abundant, a good place to search for these imprints is at the Castle Rock Lake Recreation Area. A level hiking trail follows the lakeshore and strides by several small but productive cuts of red clinker ("scoria") that contain fossils. With the help of a hammer and chisel, it should be no problem to collect a decent sample, but there are only a couple outcrops large enough to search. Of course the walk around the small quiet lake isn't so bad either.

Several abandoned "scoria" pits and tailings piles that are said to contain abundant fossils can also be found in the surrounding area, all located on private property. Check with landowners before attempting to collect or visit any of these sites.

Roundup

See map on page 182

Land type: Low mountains, road cut.

GPS: Large and sporadic area.

Best season: Any.

Land manager: U.S. Highway, maintained by Montana DOT.

Material: Fossils.

Tools: Rock hammer, chisel.

Vehicle: Any.

Accommodations: Camping, RV parking, and motels in Roundup.

Special attractions: Musselshell Valley Historical Museum in Roundup has a small selection of fossils as well as historical relics.

Finding the site: From Roundup drive south on U.S. Highway 87. The road cuts begin just outside of town and extend for 15 miles, with the cuts farthest south from Roundup often producing the best fossils. *DeLorme: Montana Atlas & Gazetteer:* Page 44 B2 and C2.

Rockhounding

Roundup was once a coal-mining town, and today the rockhound can benefit from the imprint of coal—literally. This highway leading out of Roundup has wonderfully preserved fossil prints of plants in the layers of the road cuts. The primary strata are Tertiary sandstones and shales belonging to the Fort Union Formation. These layers are easily identifiable, and the best fossil samples are obtained from splitting the layers with a hammer and chisel. Prints of plants can be collected easily above and below the coal vein but are not as finely detailed as prints from splitting the other layers. As with any delicate plant fossil, bring proper packaging equipment for the fragile prints.

Flatwillow Creek

See map on page 192

Land type: Hills, road cut.

GPS: N 46.79861 / W. 108.61194

Best season: Spring through fall.

Land manager: U.S. Highway, maintained by Montana DOT.

Material: Fossils.

Tools: Rock hammer.

Vehicle: Any.

Accommodations: Camping, RV parking, and motels in Roundup.

Special attractions: Musselshell Valley Historical Museum in Roundup has a small selection of fossils as well as historical relics.

Finding the site: From Roundup drive 24 miles north on U.S. Highway 87 to several rock outcrops along the road. Collecting areas are located in a large spread on both sides of US 87, about 3 to 5 miles past the Montana Highway 244 junction to Winnett. *DeLorme: Montana Atlas & Gazetteer:* Page 60 D1.

Rockhounding

This area can produce some interesting fossils. The first locality is encountered about 3 miles north of the MT 244 junction to Winnett. Here, as US 87 descends into the valley of Flatwillow Creek, several road cuts in exposures of the Cretaceous Colorado Shale are found. The Mosby Sandstone is well exposed in the northernmost road cut on the east side of the highway. Specimens of fossil oysters can be picked up on the surface road cut, while sandy concretions containing a fair abundance of fossil snails are found here and there. Some of the adjacent road cuts contain limestone concretions with attractive white and brown calcite crystals. Crystals of selenite are abundant on the surface of all the road cuts.

About 2 miles farther north on US 87 (5 miles north of the MT 244 junction to Winnett), are some low road cuts on both sides of the highway exposing the Cretaceous Mowry Shale. By splitting the light-colored, thin-bedded layers, one can find abundant dark brown fish scales that average ¼ inch in size. Some magnification (10X hand lens) will reveal the concentric ring pattern on the surface of the scales.

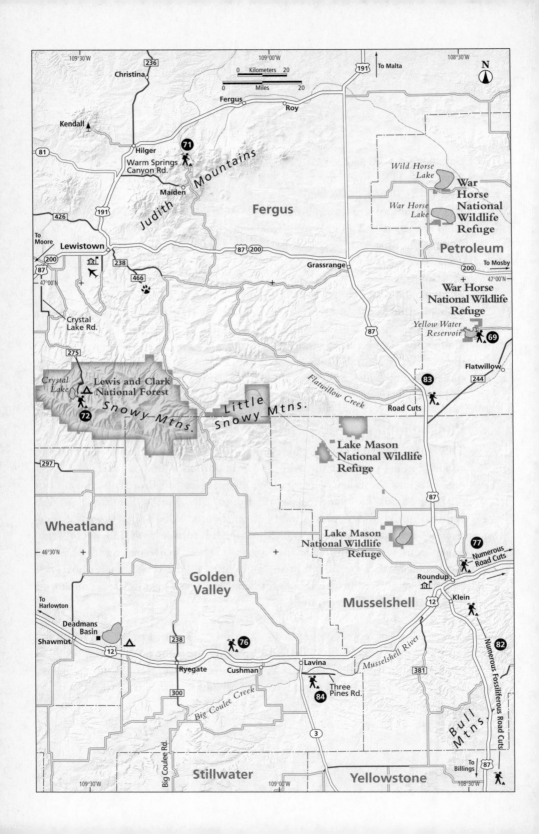

Lavina

See map on page 192
Land type: Hills, road cut.
GPS: N. 46.27804 / W. 108.93036
Best season: Any.
Land manager: Golden Valley County.
Material: Fossils.
Tools: Rock hammer.
Vehicle: Any.
Accommodations: Camping, RV parking, and motels in Roundup.
Special attractions: None.
Finding the site: From the junction of U.S. Highway 12 and Montana Highway 3 in Lavina, drive south on MT 3 for 1.8 miles. Turn left onto a dirt road

A thick 3-foot layer of solid oyster conglomerate south of Lavina.

immediately after Big Coulee Creek and drive 0.4 mile to a road cut on a small hill with a dark vein of lignite. Fossilized oysters can be found in a thick vein beneath the lignite. *DeLorme: Montana Atlas & Gazetteer:* Page 43 B9.

Rockhounding

Although this is a very small site, the 3-foot-thick chunky layer of solid fossilized Cretaceous oysters is proof that good things can come in small road cuts. This site provides plenty of collecting material and demonstrates the formation of theses rocks in a shore environment with alternating periods of terrestrial plant material and aquatic life. The lignite is a result of the drier times where terrestrial plants flourished, as lignite is a poor grade of coal and forms from these conditions. The alternating layers in the road cut of sandstone, shale, oyster conglomerate, and lignite are evidence of the changing environments of the past. The oysters found here belong to a single genus and species, *Crassostrea subtrigonalis.*

Riverfront Park

See map on page 196
Land type: River, gravel bars.
GPS: N. 45.73967 / W. 108.54126
Best season: Spring through summer.
Land manager: Montana Fish, Wildlife, and Parks.
Material: Agate, petrified wood, chert, jasper.
Tools: None.
Vehicle: Any.
Accommodations: Camping, RV parking, and motels in Billings.
Special attractions: Pictograph Cave National Historic Landmark south of Billings. "Pompey's Pillar" is a popular tourist attraction; the outcrop of Cretaceous sandstone along the Yellowstone River was signed by Captain Clark of the Lewis and Clark expedition in 1806.
Finding the site: From Interstate 90 in Billings, exit at South Billings Boulevard. Drive south for 0.7 mile and turn left into Riverfront Park. Stay right on the road inside the park, which dead-ends at several parking spaces along the Yellowstone River. *DeLorme: Montana Atlas & Gazetteer:* Page 30 A2.

Rockhounding

The gravels in and adjacent to the Yellowstone River south of Billings are a source of petrified wood and occasional agates. Banded chert and red jasper also frequent the gravels. The assorted siliceous rocks cut and polish nicely. The translucent stones can be found in several colors and patterns by simply walking the shore. Periods of low water are always the best time to search in river gravels for potential cutting material. A bike trail at the parking area follows the river. This is also an excellent area for picnicking since it's within a large green park with facilities and a small lake.

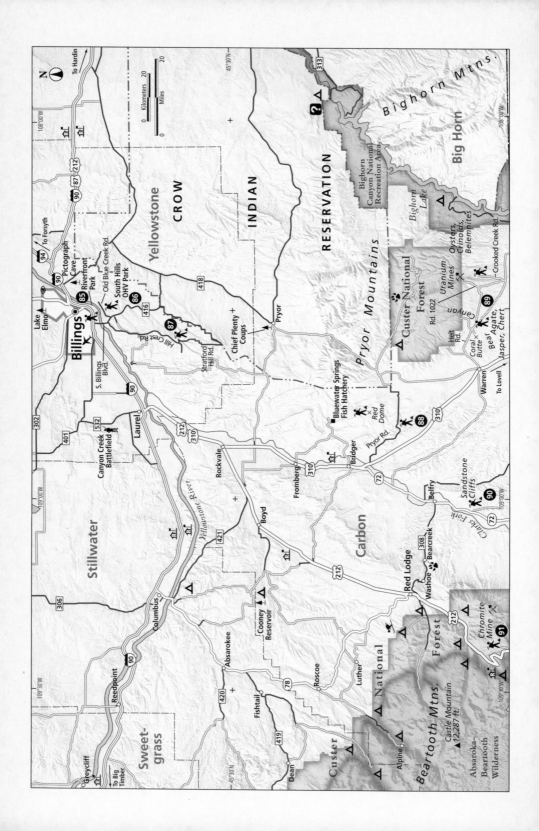

South Hills

See map on page 196
Land type: Hills.
GPS: N. 45.73580 / W. 108.52523
Best season: Spring through fall.
Land manager: Bureau of Land Management (BLM).
Material: Calcite crystals, barite crystals, celestite crystals, fossils.
Tools: Rock hammer.
Vehicle: Any.
Accommodations: Camping, RV parking, and motels in Billings.
Special attractions: Pictograph Cave National Historic Landmark south of Billings. "Pompey's Pillar" is a popular tourist attraction; the outcrop of Cretaceous sandstone along the Yellowstone River was signed by Captain Clark of the Lewis and Clark expedition in 1806.
Finding the site: From Interstate 90 in Billings, exit at South Billings Boulevard and drive south for 1.2 miles. Turn left onto Old Blue Creek Road and drive 0.4 mile to the South Hills OHV Park and park. *DeLorme: Montana Atlas & Gazetteer:* Page 30 A2.

Rockhounding

The Cretaceous Colorado Shale of marine origin makes up the high hills along the Yellowstone River in this area directly south of Billings. Throughout the formation numerous limestone concretions can be found, most of which are septarian in nature, with veins of white and brown calcite. Periodically, crystals of barite and celestite are found associated with calcite crystals in small cavities within the veins themselves. Fossils, too, are sometimes found in some of the concretions, and although pelecypods predominate, some relatively nice specimens of ammonites have been found.

Search the area in and around the base of the hills. Crystals can be found by either splitting chunks of limestone or searching the loose chunks at the base of the hills. Fossils are not very common. It should be noted that although this park is open for collecting crystals and fossils, it is designated as an off-road vehicle park for motorcycles and other ATVs. The rockhound should stay in open areas and be cautious of the off-road traffic; riders may neither hear nor see the presence of foot traffic. Exercise caution, especially with children. The rockhound is a guest in this area.

Stratford Hill

See map on page 196
Land type: Hills, road cut.
GPS: N. 45.58413 / W. 108.57943
Best season: Any.
Land manager: Yellowstone County.
Material: Fossils.
Tools: Rockhammer, chisel, goggles.
Vehicle: Any.
Accommodations: Camping, RV parking, and motels in Billings.
Special attractions: Pictograph Cave National Historic Landmark south of Billings. "Pompey's Pillar" is a popular tourist attraction; the outcrop of Cretaceous sandstone along the Yellowstone River was signed by Captain Clark of the Lewis and Clark expedition in 1806.

An ammonite imprint from Stratford Hill Road.

Finding the site: From Interstate 90 in Billings, exit at South Billings Boulevard. Drive 1.7 miles south and turn right onto Hill Crest Road. Continue 4.3 miles on Hill Crest Road until it becomes gravel and forks. Stay left, drive 5.7 miles, and turn left onto Stratford Hill Road. Continue for 2 miles to a large road cut in the highest hill on the left side of the road. *DeLorme: Montana Atlas & Gazetteer:* Page 30 B2.

Rockhounding

The Mowry Shale at this site contains flattened specimens and imprints of ammonites and extremely well-preserved fish scales, some of which retain their iridescence and colors. Located in the rolling hills just outside of Billings, the 20-foot-tall layers of the road cut can be tempting to explore all day, especially since the rock is easily split and quite productive. Fossils also can be found by looking through the loose chunks at the bottom of the cut. The best specimens have been found by carefully splitting the larger pieces of the shale with a rock hammer and a good pair of goggles. The fish scales are abundant; the ammonites and imprints, especially a whole one, require a more diligent search.

Red Dome

See map on page 196
Land type: High hills, road cut.
GPS: N. 45.20663 / W. 108.79948
Best season: Spring through fall.
Land manager: Bureau of Land Management (BLM).
Material: Fossils, petrified wood.
Tools: Rock hammer.
Vehicle: Any; four-wheel drive when wet.
Accommodations: Camping, RV parking, and motels in Red Lodge.
Special attractions: Chief Plenty Coups National Historic Monument at the Crow Indian Reservation includes a museum of Native American history. South of the site the Pryor Mountain Wild Horse Range offers an opportunity to view wild horses.
Finding the site: From Bridger drive south on U.S. Highway 310 for 2.6 miles, turn left onto Pryor Road, and continue 8.6 miles to a collecting area. A BLM

Oyster conglomerate from Red Dome.

map is recommended to ensure no trespassing, because the road traverses through private and public land. The area around Red Dome produces a variety of fossils; the site listed is a recommended starting point. *DeLorme: Montana Atlas & Gazetteer:* Page 30 C1.

Rockhounding

Red Dome itself is an interesting geologic formation, and the area has been the subject for unique finds, including several species of dinosaurs. The dome is an arch of Triassic mudstones, an anticline with a striking red color that dominates the area visually.

There are several localities both on private land and BLM land east of Bridger that show excellent exposures in the marine Jurassic Piper, Rierdon, and Swift Formations. These rock units contain a relative abundance of fossils, including amazing oyster conglomerates, belemnites, star crinoids columnals, and associated fauna, including several species of clams, snails, and ammonites. At the particular location listed, Red Dome rests as a phenomenal backdrop to the road cut with marine fossils. Look for chunks of limestone oyster conglomerates loose on the hill on the right side of the road. My first sample contained about 200 oysters in a hand-held piece. The GPS mentioned should simply be a starting place. With a BLM map there are many places to explore in this area.

Pryor Mountains

See map on page 196
Land type: Hills, badlands.
GPS: N. 45.03700 / W. 108.49644
Best season: Spring through fall.
Land manager: Bureau of Land Management (BLM).
Material: Fossils, agate, jasper, chert, pyrite, calcite crystals, fluorite, tyuyamu-nite.
Tools: None.
Vehicle: Any.
Accommodations: Camping on-site at Bear Canyon Road; RV parking, and motels in Lovell, Wyoming.
Special attractions: Chief Plenty Coups National Historic Monument at the Crow Indian Reservation includes a museum of Native American history. South of the site the Pryor Mountain Wild Horse Range offers an opportunity to view wild horses.
Finding the site: From U.S. Highway 310 in Warren, turn left onto the road leading toward the quarry. Drive 2.7 miles and turn right onto Helt Road. The first site is a large red butte 3.4 miles down Helt Road, across from Bear Canyon Road; a road goes right toward the base. Park at the bottom and explore the butte for fossilized coral.

The next site can be reached by continuing east on Helt Road for 1 mile to a fork in the road. Park here and explore the roadside and surrounding hills for loose fossils, agate, chert, and jasper.

The next site can be reached by going left at this fork (onto Helt Road), driving 5.2 miles, and turning left onto BLM Road 1022. Drive 1 mile to the uranium mines. (This option is for high-clearance four-wheel-drive vehicles only.)

The last site can be reached by driving 4 miles past Road 1022 farther down Helt Road and turning left onto Crooked Creek Road. A nice area to begin collecting star crinoids and oysters can be found 2.7 miles down Crooked Creek Road. Search the surrounding hills and wash beds for loose fossils. Loose shale at the tops of the hills can be split easily, exposing the occasional insect. *DeLorme: Montana Atlas & Gazetteer:* Page 30 D2.

(From top left to right to bottom left to right) Pentacrinus columnals, the oyster Gryphaea, the coral Astrocoenia, the bullet-shaped belemnite Pachyteuthis, and samples of banded chert and jasper from exposures of marine Jurassic rocks in the Pryor Mountain region.

Rockhounding

The Pryor Mountains consist of several fault blocks that were uplifted by diastrophic processes during the Late Cretaceous and Early Tertiary periods. As a result of this mountain building and subsequent erosion, sedimentary rock layers ranging in age from Early Paleozoic to Late Cretaceous occur within and on the margins of the range. Almost all of the rock layers are fossiliferous to varying degrees, and some contain unusual mineral deposits and rocks with lapidary potential.

The Madison limestone, a formation of late Paleozoic age, reveals spirifer brachiopods, coral, and crinoid remains massed together in some outcrops. Collapsed caverns within the limestone have been the source of some secondary uranium minerals. The low-grade ore was mined on a small scale in the

David Feldman and Jimmy Goodman split shale in the Pryor Mountains.

past. The dumps of these mines will yield nice specimens of calcite crystals, small fluorite crystals, and pyrite. The uranium mineral, tyuyamunite, occurs as a yellow powdery coating on the calcite crystals or as disseminated grains throughout the brecciate limestone. Exposures of Late Paleozoic sediments on the flanks of the Pryors contain an abundance of limonite after pyrite or goethite after pyrite pseudomorphs. These occur as clusters of cubic crystals. Most are deeply oxidized, but some retain a very dark color with submetallic luster. Many of them, if broken, reveal original pyrite in the center.

Along the western and southern margins of the Pryors are excellent exposures of Mesozoic rocks. The bright red Triassic Chugwater Formation is quite noticeable in the landscape but is nonfossiliferous. Marine Jurassic formations directly overlaying the Chugwater Formation, however, provide an abundance of fossils. At this site the colonial coral Astrocoenia is found. The belemnite Pachyteuthis, oysters of the genera Gryphaea and Ostrea, and the star-shaped columnals of the crinoid Pentacrinus are especially prolific.

The Pryor Mountains are a personal favorite because there are so many treasures to be found and such interesting stratigraphy to keep the geologic mind intrigued. Throughout the area is an abundance of fossils, jasper, and agate that can be found easily by exploring the ground, particularly the wash beds of the area, along with the intact layers of the hills. Your best bet is to camp on-site at the BLM Bear Gulch Campground and spend a couple of days wandering this beautiful high desert.

Belfry

See map on page 196
Land type: Hills.
GPS: N. 45.06237 / W. 109.03893
Best season: Spring through fall.
Land manager: Bureau of Land Management (BLM).
Material: Fossils, petrified wood.
Tools: Rock hammer.
Vehicle: Any.
Accommodations: Camping, RV parking, and motels in Red Lodge and Powell, Wyoming.
Special attractions: None.
Finding the site: From the intersection of Montana Highways 72/308 in Belfry, drive south on MT 72 for 5.8 miles. Turn left 0.7 mile after crossing the Clarks Fork of the Yellowstone River onto a faint dirt road leading up the side of the hills on the BLM land. Drive just a couple hundred feet up the road to sandstone containing the fossils or park at the bottom and walk. *DeLorme: Montana Atlas & Gazetteer:* Page 29 D8.

Rockhounding

Abundant fossils of freshwater clams can be found in the Fort Union Formation south of Belfry just before the Wyoming state line. Most of the fossils are highly weathered, but well preserved complete clams with both side of the shell can be found loose from the parent rock. The large specimens are found mostly in the thick slabs of sandstone that is almost impossible to break. With diligent searching quality specimens can be found loose or in smaller chunks of rock. The concentration of exposed sandstone that contains the fossils seems to be limited solely to the area mentioned, although a couple hundred feet farther straight up the hill samples of petrified wood can be found. The wood is found loose on the ground, is usually a few inches long, and occurs in rust-colored shades.

Beartooth Chromite Mine

See map on page 196
Land type: High plateau.
GPS: N. 45.03278 / W. 109.40748
Best season: Summer.
Land manager: USDA Forest Service, Custer National Forest.
Material: Chromite, serpentine, feldspar crystals.
Tools: Rock hammer.
Vehicle: Any.
Accommodations: Camping and RV parking in Custer National Forest; camping, RV parking, and motels in Red Lodge and Cooke City.
Special attractions: Yellowstone National Park.
Finding the site: From Red Lodge drive south on U.S. Highway 212, the Beartooth Scenic Highway, for 11.6 miles to the top of the Line Creek Plateau and park at the scenic overlook on the right side of the road. The chromite mine is located on the side of the road opposite the parking area; use caution crossing the highway. Follow the old jeep trail on the top of the knoll across from the parking area. Motorized travel is restricted, and a hike of about 0.5 mile is required to reach the site along the jeep trail. It should also be mentioned that since the altitude of the site is close to 10,000 feet, one can tire quickly from a small amount of exertion. This site should be visited only by those in good health and children old enough to hike in such high altitude. The chromite deposits here can be reached only during the summer months because of snow, and storms have been known to deposit significant depths of snow even during the summer months. *DeLorme: Montana Atlas & Gazetteer:* Page 29 D7.

Rockhounding

Several unique mineral deposits occur within the highly deformed metamorphic rocks that make up the core of the Beartooth Mountains in southern Montana. Many of these deposits were mined for chromite during the 1930s and 1940s when foreign supplies were not readily available. Chromite is the major ore mineral of chromium, a metal that has many uses, the most important being that of strengthening steel. Though not mined as extensively as the deposits within the Stillwater drainage to the north, the deposits located south of Red Lodge on Line Creek Plateau can produce samples of chromite and other interesting mineral specimens.

Striking views from above the tree line on the Beartooth Pass.

Rocks in the area of the mine contain an assortment of minerals. The host rock, a dark green serpentine (locally called "Montana jade"), contains the chromite. The serpentine can be fashioned by those who are skilled in the art of lapidary, but inclusions of other minerals prevent it from being a quality gemstone. The chromite is generally black and granular. Often, weathered serpentine between the grains of chromite imparts a gray appearance to the ore.

Perhaps the most interesting mineral here is orthoclase feldspar, which can be found as relatively large crystals weathering from the light-colored rocks. These light-colored rocks are part of a small igneous intrusion, much younger than the Precambrian metamorphic rocks containing the chromite. The crystals are hexagonally shaped and range in size from ¼ to more than 1 inch long. Most of the crystals loose in the soil have been partially altered to kaolinite. Better crystals must be carefully broken from the weathered rock containing them.

Appendix A: Glossary

Agate: A variety of chalcedony closely related to opal that is often translucent and found in a variety of colors and patterns.

Ammonite: An extinct group of mollusks that may have been similar to the present–day chambered nautilus.

Amethyst: A purple or violet variety of quartz often used as a gem.

Anticline: An upturned fold in the rocks of the earth's crust.

Aquamarine: A blue–green gem variety of beryl.

Azurite: A blue copper carbonate often associated with malachite.

Baculite: An extinct cephalopod that's similar to ammonites but is straight rather than curled.

Barite: Barium sulfate occurring in blue, green, brown, and red colors.

Batholith: A large mass of plutonic rock many tens of square miles in area.

Belemnite: An extinct cephalopod possessing a cigar–shaped protective guard.

Beryl: Beryllium aluminum sulphate that is generally colorless in its pure form; varieties include blues, pinks, greens, and yellows. Beryl constitutes such precious gems as emerald and aquamarine.

Brachiopod: A marine animal with two unequal shells showing bilateral symmetry.

Breccia: Angular rock fragments cemented into solid rock.

Bryozoans: Small colonial animals that build calcareous structures in which they live.

Cabochon: A gemstone that has been fashioned into a dome and polished.

Calcite: A common crystalline form of calcium carbonate that is usually white or gray.

Cephalopods: A class of mollusks that includes the squid, the octopus, and the chambered nautilus.

Chalcedony: A microcrystalline variety of quartz.

Chert: An extremely fine–grained siliceous rock exhibiting many colors.

Columnal: A portion of the column, or "stem," of crinoids.

Concretion: A nodular lumpy rock, generally sedimentary, that forms about a nucleus (often a fossil).

Country Rock: The common rock surrounding another deposit of material.

Crinoid: A "flowerlike" echinoderm with a multi–armed calyx, or head, and a long column, or "stem," attaching it to the seafloor.

Dendrite: A tree–like pattern produced when minerals (usually oxides of manganese) crystallize in minute fractures in the rocks.

Diastrophic: Natural processes that deform the earth's crust (tension and compression that produce faults, folds, etc.).

Dike: An igneous intrusion that cuts across pre–existing rock layers.

Echinoderm: A phylum of marine invertebrates with spiny bodies, including starfishes, sea urchins, and sea cucumbers.

Echinoids: A class of echinoderms that includes the sand dollar and sea urchin.

Epidote: A green monoclinic mineral with crystals that are often used as gemstones.

Fault: A break or fracture in the earth's crust along which movement takes place.

Feldspar: The most widespread mineral group occurring in shades of white, gray, and pink in all kinds of rock.

Fossil: The remains of plants or animals preserved in rocks.

Fossiliferous: Containing fossils.

Fluorite: A clear–to–translucent mineral commonly occurring in shades of blue or purple as a common mineral in veins, which can be found in cubic crystals.

Garnet: Any mineral of the garnet group, commonly used as an abrasive. Garnet also occurs as a pink–red semiprecious gem commonly cut into stones.

Gastropod: A type of mollusk with an asymmetrical unchambered shell.

Gem: A general term for a variety of stones that can be cut for ornamental purposes.

Geode: A hollow nodule or concretion that may contain crystals.

Gypsum: A widely distributed mineral consisting of hydrous calcium sulphate that commonly forms in thick beds and occurs in shades of transparent (selenite), red, brown, and gray.

Genus: One of the divisions in the classification of living things (or fossils).

Geology: The science that studies the earth, its composition, the processes that affect the rocks of which it is composed, and its history.

Geomorphology: The geologic study that deals with the shape of the earth's surface and the development of landforms.

Gradation: A natural process such as weathering, erosion, or deposition that helps to shape the surface of the earth.

Hydrothermal: Processes involving the action of hot–water solutions.

Igneous: Rock that hardened from a molten state.

Intrusion: Igneous rock that was intruded into pre–existing rocks.

Jasper: A variety of chert associated with iron ore that typically occurs in the color of red.

Laccolith: An igneous intrusion that has squeezed between older rock layers and has domed up the overlying strata.

Lapidary: The art or artist who fashions gemstones from rough rock.

Limonite: A general term for a group of brownish iron hydroxide, commonly a secondary mineral due to oxidation of iron–bearing minerals.

Malachite: A green copper ore that occurs in crystal and massive forms.

Metamorphic: Rocks that have undergone physical and chemical change due to extreme changes in temperature and pressure.

Mica: A group of sheet silicate minerals, major members of which include muscovite and biotite.

Micromount: Small mineral specimens that have been permanently mounted and are best observed with the aid of a microscope.

Mollusk: Any of the marine invertebrates in the phylum Mollusca, generally characterized by soft unsegmented bodies and hard calcareous shells, which includes cephalopods, pelecypods, gastropods, etc.

Opal: A silicon oxide closely related to chalcedony. It occurs in a dull common form and the translucent precious form, which may contain flashes of "fire" and is used as a gemstone.

Paleontology: The branch of science that deals with the study of fossils.

Pegmatite: An exceptionally course–grained igneous rock with interlocking crystals usually occurring at the edges of batholiths; also found as lenses and veins.

Pelecypods: A class of bivalve mollusks that includes oysters and clams.

Placer: Generally a sand or gravel deposit containing minerals of economic value.

Plutonic: Igneous rocks that hardened deep within the crust of the earth.

Porphyry: An igneous rock with crystals of different sizes—usually large crystals surrounded by very small crystals.

Pseudomorph: A crystal with the geometric form of a mineral that has been replaced chemically by another mineral.

Pyrite: Metal–looking sulfides or disulfides. The brassy form is often known as "fool's gold."

Quartz: Common rock–forming mineral composed of silicon dioxide. Crystals may be glassy or opaque (milky quartz) and exist in a variety of colors, including white, rose, smoky gray, and purple.

Sedimentary: Rocks that form from the accumulation, compaction, and cementation of sediment.

Septarian Concretion: A concretion possessing internal fractures that are filled or partially filled with minerals.

Siliceous: Containing silica.

Sill: An igneous intrusion that parallels pre–existing rock layers.

Spirifer: A type of brachiopod possessing a butterfly–shaped shell.

Stratigraphy: The scientific study of the layers of strata in the earth's crust.

Syncline: A down–turned folds in the rocks of the earth's crust.

Thumbnail Specimen: A small mineral specimen about 1 inch by 1 inch in size.

Trilobite: Extinct marine arthropod that is beetle–like in appearance.

Volcanic: Igneous rocks that hardened on the earth's surface.

Vug: A small cavity in a rock.

Zeolite: A general term for a group of hydrous aluminosilicates that can occur as well–formed crystals in cavities of basalt.

Appendix B: Further Reading

Alt, David, and Donald W. Hyndman. *Roadside Geology of Montana.* Missoula, MT: Mountain Press Publishing Co., 1986.

Bates, Robert L., and Julia A. Jackson. *Dictionary of Geologic Terms.* New York: Random House, Inc., 1983.

Harmon, Tom. *The River Runs North: A Story of Montana Moss Agate.* Sidney, MT: Elk River Printing, 2000.

Korbel, Petr, and Milan Novak. *The Complete Encyclopedia of Minerals.* Edison, NJ: Chartwell Books, Inc., 1999.

Montana Bureau of Mines and Geology. Various bulletins (www.mbmg .mtech.edu).

Prinz, Martin, George Harlow, and Joseph Peters. *Simon and Schuster's Guide to Rocks and Minerals.* New York: Simon and Schuster, Inc., 1978.

Thompson, Ida. *National Audubon Society Field Guide to North American Fossils.* New York: Alfred A. Knopf, Inc., 2001.

Appendix C: Finding Maps, Land Managers, and Information

USDA Forest Service

Beaverhead–Deerlodge National Forest
Forest Supervisor's Office
Thomas Reilly, Forest Supervisor
420 Barrett Street
Dillon, MT 59725–3572
(406) 683–3900 or (406) 683–3913

Bitterroot National Forest
Forest Supervisor's Office
1801 North 1st Street
Hamilton, MT 59840–3114
(406) 363–7100 or (406) 363–7100

Custer National Forest
Supervisor's Office
1310 Main Street
Billings, MT 59105
(406) 657–6200

Flathead National Forest
Supervisors Office
1935 3rd Avenue East
Kalispell, MT 59901
(406) 758–5200

Gallatin National Forest
Supervisor's Office
P.O. Box 130
Bozeman, MT 59771
(406) 587–6701 or (406) 522–2520

Helena National Forest
Supervisor's Office
2880 Skyway Drive
Helena, MT 59602
(406) 449–5201

Kootenai National Forest
Supervisor's Office
1101 Highway 2 West
Libby, MT 59923
(406) 293–6211

Lewis and Clark National Forest
Supervisor's Office
1101 15th Street North
Great Falls, MT 59405
(406) 791–7700

Lolo National Forest
Supervisor's Office
Fort Missoula Building 24
Missoula, MT 59804
(406) 329–3750

Bureau of Land Management

Billings Field Office
5001 Southgate Drive
Billings, MT 59107
(406) 896–5013

Butte Field Office
106 North Parkmont
Butte, MT 59701
(406) 533–7600

Dillon Field Office
1005 Selway Drive
Dillon, MT 59725
(406) 683–2337

Glasgow Field Office
Highway 2 West
RR #1 – 4775
Glasgow, MT 59230
(406) 228–3750

Lewistown Field Office
Airport Road
P.O. Box 1160
Lewistown, MT 59457
(406) 538–7461

Malta Field Office
HC 65, Box 5000
Malta, MT 59538
(406) 654–1240

Miles City Field Office
111 Garryowen Road
Miles City, MT 59301
(406) 233–2800

Missoula Field Office
3255 Fort Missoula Road
Missoula, MT 59801
(406) 329–3914

Montana Bureau of Mines and Geology

Montana State University–Billings
1300 North 27th Street
Billings, MT 59101
(406) 657–2938

Montana Tech
1300 West Park Street
Butte, MT 59701
(406) 496–4167

US Geological Survey

Western Distribution Branch
Box 25286, Denver Federal Center
Denver, CO 80225
(303) 202–4700
(sells Montana USGS maps and publications)

Appendix D: Rock Shops in Montana

A&L Shoppers Pawn & Rock Shop
2101 Harrison Avenue
Butte, MT 59701
(406) 782–9000

The Agate Stop—Home of the
Montana Agate Museum
124 Fourth Avenue North
Savage, MT 59262
(406) 776–2373

Crystal Hound
160 8th Avenue West North
Kalispell, MT 59901
(406) 756–2357 or (406) 212–2602

Crystal Imports
500 South Russell Street
Missoula, MT 59801
(406) 549–8907

Crystal Limit
2301 South Grant Street
Missoula, MT 59801
(406) 549–1729

Earth's Treasures
25 North Wilson Avenue
Bozeman, MT 59715
(406) 586–3451

Gold Miser
30525 U.S. Highway 2 South
Libby, MT 59923
(406) 293–8679
(prospecting supplies only)

Healing Wonders
442 Batavia Lane
Kalispell, MT 59901
(406) 756–8705

Junction Rock Shop
7309 U.S. Highway 2 East
Columbia Falls, MT 59912
(406) 756–8705

Kehoes Agate Shop
1020 Holt Drive
Big Fork, MT 59911
(406) 837–4467
(they sell rocks, too)

Montana Within Rock Shop
1761 Columbia Falls Stage
Columbia Falls, MT 59912
(406) 755–4788

The Prospector Shop
6312 U.S. Highway 12 West
Helena, MT 59601
(406) 442–1872

Rocks and Fossils
5170 U.S. Highway 89 South
Livingston, MT 59047
(406) 222–6725
(Also inquire about the new
location—Dancing Bear Gallery of
Gems, Minerals, Fossils, Jewelry and
Beads Shop in Bozeman)

Index

Agate, 8–10, 56, 57, 76, 79, 80–82, 83, 166, 176, 188, 195, 202, 205
Amethyst, 47–48, 139
Andesite, 146
Aquamarine, 101, 103
Arsenopyrite, 41, 86–87, 88, 93, 127–29
Augite, 47–48
Azurite, 55, 73, 75, 84, 85, 127–29

Barite crystals, 49, 89, 177, 197
Belt rock, 25, 26, 32, 46, 58, 66, 69, 130

Calcite, 38, 47–48, 61–62, 101, 103, 114, 142, 143, 149, 153, 155, 159, 160, 161, 197, 202, 204
Celesite, 197
Chalcedony, 9, 56, 176
Chalcopyrite, 55, 88, 132, 151
Chert, 114, 195, 202
Chromite, 207–8
Chrysocolla, 127–29, 132
Citrine, 76, 79
Copper, 41, 50, 55, 84, 133, 152
Coral, horn, 25, 38, 116

Epidote crystals, 101, 103

Feldspar, 65, 97–98, 99, 125, 126, 169, 207, 208
Flourite, 66–67, 71, 72, 202, 204
Forsterite crystals, 134
Fossils
    dinosaur, 20, 35, 156–57, 163, 180, 201

marine, 23, 25, 30–31, 37, 38, 49, 59, 116, 118–19, 153, 155, 159, 160–63, 171–72, 175, 179, 183, 191, 193, 197, 198, 199, 200, 201, 202–5, 206
    plant, 39, 63, 138, 139, 147, 148, 178–79, 184, 187, 189, 190

Galena, 33, 34, 55, 73, 75, 88, 93, 127–29
Garnet, 27, 76, 78, 79, 97, 101, 103, 122–24, 125, 126, 151, 158
Gold, 18, 27, 29, 32, 34, 41, 50, 55, 69, 73–75, 76, 88, 89, 94, 119, 127, 130, 158, 159, 173–75
Granite, 32, 65, 69, 99
Gypsum, 116

Heulandite, 142, 143
Hyalite, 144, 145

Iron, 185

Jasper, 55, 56, 57, 76, 83, 146, 166, 176, 188, 195, 202, 205
Jasperoid, 114
Jurassic coal, 39

Kaolinite, 208

Laumontite, 47–48, 142–43
Lead, 41, 50, 55, 114, 133
Lignite, 194
Limestone, 25, 38, 59, 84, 116, 118–19, 127, 133, 150, 159, 171, 172, 179, 191, 201, 203, 204

Limonite, 44, 46, 204

Magnetite, 151, 152
Malachite, 50, 55, 84, 85, 88, 114, 127–29, 132, 133
Manganese, 94
Mesolite, 47–48
Mica, 99, 127–29
Moonstone, 76, 79
Mudcracks, Precambrian, 26, 91–92
Mudstone, 26, 201

Opal, 136, 145

Parisite, 66, 68
Petrified wood, 19, 134, 138, 139–141, 166, 176, 177, 185, 188, 195, 200, 206
Precambrian rock, 29
Psilomelane, 84
Pyrite, 41, 44, 46, 50, 55, 58, 73, 75, 84, 85, 88, 93, 114, 127–29, 151, 202, 204
Pyrolusite, 93, 94
Pyrrohite, 33, 34

Quartz, 41, 55, 64–65, 66, 67–68, 73, 75, 76, 79, 82, 84, 85, 87, 88, 97–98, 99–100, 104, 106–7, 109–10, 125, 126, 127–29, 139, 146, 151, 168, 169–70

Raindrop imprints, 91–92
Rhodochrosite, 93, 94
Rhyolite, 87, 136
Ripple marks, 26, 91–92
Ruby, 76, 79

Sandstone, 20, 37, 49, 148, 164, 177, 178–79, 184, 191, 206
Sapphire, 8, 9, 76–79, 95–96, 130, 173–75
Scheelite, 101
Schorl, 97–98
Selenite, 177, 179, 191
Serpentine, 207, 208
Shale, 20, 30–31, 38, 39, 49, 59, 63, 84, 155, 161, 163, 164, 171, 178, 179, 184, 187, 191
Silver, 41, 50, 55, 93, 94, 114, 133, 158
Soapstone, 111, 113
Sphalerite, 33, 34, 55, 73, 75, 88, 93, 127–29
Stibnite, 33, 34
Stilbite, 47–48, 52, 54
Sulphur, 50
Sulphur coal, 189

Tailings, 27, 34, 41, 50, 55, 75, 85, 88, 94, 110, 115, 119, 127, 151–52, 189
Tetrahedrite, 93
Topaz, 76, 79, 86–87
Travertine, 150
Tyuyamunite, 202

Uranium minerals, 203–4

Vermiculite, 29
Volcanic rock, 54, 87, 88, 136

Wonderstone, 120–21

Zinc, 41, 50, 133

About the Authors

Montana Hodges was born in California and was named Montana by her parents for no particular reason. As a graduate of journalism and geology, she is seriously committed to summer rockhounding trips. Today she works as a free-lance writer and travels between her home in California and wherever the rocks may take her.

Robert Feldman is a retired public-school earth science teacher who lives with his wife, Claudia, in Billings, Montana. Bob has three grown children, Barb, Tom, and David, who have frequently accompanied him on fossil-collecting outings. He developed an interest in invertebrate paleontology at an early age and spends considerable time studying the subject. Along with his family, he has explored Montana's mining districts and fossil fields, intent on learning as much as he can about Montana's past.